Hidden Kingdom
The Insect Life of Costa Rica

PIOTR NASKRECKI

Foreword by Edward O. Wilson

A Zona Tropical Publication

FROM

Comstock Publishing Associates

a division of

Cornell University Press

Ithaca and London

First published 2017 by Cornell University Press

Printed in China

Library of Congress Cataloging-in-Publication Data

Names: Naskrecki, Piotr, author. | Wilson, Edward O., writer of foreword.
Title: Hidden kingdom : the insect life of Costa Rica / Piotr Naskrecki ; foreword by Edward O. Wilson.
Description: Ithaca : Comstock Publishing Associates, a division of Cornell University Press, 2017. |
    "A Zona Tropical publication." | Includes index.
Identifiers: LCCN 2017019538 (print) | LCCN 2017022930 (ebook) | ISBN 9781501709623 (pdf) |
    ISBN 9781501704710 (pbk. : alk. paper)
Subjects: LCSH: Insects—Costa Rica. | Insects—Costa Rica—Identification.
Classification: LCC QL478.C8 (ebook) | LCC QL478.C8 N37 2017 (print) | DDC 595.7097286—dc23
LC record available at https://lccn.loc.gov/2017019538

Zona Tropical Press ISBN 978-0-9894408-6-8

Book design: Gabriela Wattson and Piotr Naskrecki

This book is dedicated to the people of Costa Rica, whose work to protect and promote biodiversity sets an example for the entire world.

Este libro está dedicado al pueblo de Costa Rica, cuyo trabajo para proteger y promover la biodiversidad es un ejemplo para el mundo entero.

Contents

Foreword

I have been an entomologist for most of my long professional life, first as a student, then as researcher, teacher, and museum curator at Harvard University. So I claim some weight when I recommend this work by Piotr Naskrecki, *Hidden Kingdom,* as one of the most insightful and artistically accomplished works on insect life ever published in the literature of natural history.

The insects selected by the author (and with them other arthropods, including spiders, scorpions, and whipscorpions) are portrayed for the most part exactly as they live in the forests of Costa Rica—active in their specialized niches, secure in their micro-environments, and forever alien to human experience. If you have not seen insects as Naskrecki portrays them, then you have not experienced the heart of entomology. It takes a remarkable talent combined with drive and an intensity of effort to ferret out each of the species portrayed, then capture them photographically within the niches they occupy, with lighting that is both accurate and beautiful.

These images and Naskrecki's commentaries are notable by themselves, but *Hidden Kingdom* makes two further contributions. The first is to deepen our overall conception of what constitutes nature and wildlife. The popular definition of nature is that it is the living part of the wild environment, comprising the local flora and fauna. For most readers, flora denotes trees, shrubs, and flowering plants. Fauna means mostly if not entirely the five major groups of vertebrates: mammals, birds, reptiles, amphibians, and fishes. A few kinds of insects, including butterflies, army ants, leaf-cutter ants, and mound-building termites, often make it onto the list, but more as special guests than as frontline performers. One does not visit Yellowstone to see mountain butterflies or the Serengeti to view termite colonies.

The disproportion is magnified when one recalls that 63,000 species of vertebrates are known worldwide, compared with over a million insect species. Close attention to the "hidden kingdom" rights that imbalance. Naskrecki not only guides us into the homes of these majority creatures, he also provides a running commentary on the startling, often bizarre, details of their private lives. Insect behavior is strange enough to ordinary human experience as to suggest what life might be like on a distant planet. That would be riveting. As such, life here on Earth, in the Costa Rican rainforest, deserves our attention.

Which brings me to the second contribution. This work on insects might immediately serve a broader purpose, as an introduction to science itself. In general young people are presented with two broad routes into science. The first is STEM (science, technology, engineering, and mathematics). If a student has a special interest in one part of this quartet and is willing to work hard, he or she will find support and help to go forward in a STEM-centered career. The student may then come across the world of mystery awaiting exploration in ecosystems and the countless species composing them. The second avenue, which I happened to have traveled, is the opposite of STEM. It comes from first loving the outdoors, and then taking a scientific approach to natural history. In other words, the enthusiast plunges into the living world, then picks up STEM in order to explore nature as a branch of science. Piotr Naskrecki's *Hidden Kingdom* offers one enticing portal to this style of education and subsequent career development.

Edward O. Wilson
Harvard University
20 March 2017

Acknowledgments

Many people have helped in the preparation of this book. I would like to thank the experts who assisted with the identification of insects: Matt Buffington, Lourdes Chamorro, Stefan Cover, Lec Dayer, Geert Goemans, Fabian Haas, Akito Y. Kawahara, Petr Kocarek, Daniel Kronauer, Jack Longino, Kenji Nishida, Josip Skejo, and Doug Yanega. Some images in this book have been made possible thanks to my participation in Project ALAS at La Selva Biological Station, and I would like to thank Jack Longino and Robert K. Colwell for making me a part of it. I am indebted to the former parataxonomists of ALAS, whose help was indispensable: Maylin Paniagua, Danilo Brenes, Ronald Vargas, and Flor Cascante. I am also indebted to Bryna Belisle of the Monteverde Butterfly Gardens for allowing me to photograph butterflies there. The staff of Bat Jungle in Monteverde allowed me to photograph some of their bats. I thank Kristin Smith for her help and support during the years leading to the preparation of this book. Special thanks are owed to Kenji Nishida, a frequent travel companion in Costa Rica, a fountain of knowledge of all things entomological, and the best cook that I have ever met. I also thank Jen Guyton for her help during different stages of this book's preparation, including technical assistance with the photography, text editing, and fruitful discussions of the ideas appearing here. And finally, I am grateful to John McCuen of Zona Tropical Press for initiating—and inviting me to work on—this thoroughly enjoyable project.

Introduction

As you leave the air-conditioned environment of the airport and step onto Costa Rican soil, several things will challenge your senses. First, your skin and lungs will feel the wonderfully warm and humid atmosphere of this tropical place. And just seconds later you will realize that the air is filled with sounds—traffic noises, distorted waves of *merengue* music blaring from somebody's speaker, and, beneath all this, the steady hum of cicadas, the melodious chirping of crickets, and the mechanical buzz of a conehead katydid. Then, as you swat the first mosquito, you will notice insects in the air—among them, butterflies, dragonflies, and small beetles. Glance back at the airport building and you might notice a strange moth sitting on the wall. The realization may come to you right then and there, or a day later, but there is no escaping it—Costa Rica is a country of insects. Websites and books might have led you to believe that it is all about toucans and tapirs, but the truth is that insects are the dominant and most often seen elements of this country's diverse animal life. And this is a cause for celebration, as there is no more fascinating, beneficial, diverse, and breathtakingly beautiful group of organisms than insects.

Insects abound in every piece of habitable land and body of water. They are missing only from oceanic depths, where their close relatives, the crustaceans, play a similar ecological role. Everywhere else, insects rule. Over a million species of insects have been described by entomologists, but this is likely to represent but a fraction of the real number. Many parts of the world, especially areas in the tropics, have yet to be fully explored, and it is likely that at least 5 million more species of insects (some say 30 million) await discovery. Scientists at the Instituto Nacional de Biodiversidad (INBio) of Costa Rica estimate that at least 300,000 insect species are present in this country.

Army ant queen and worker
(*Eciton burchelli*).

Clearwing butterflies (*Hypoleria cassotis*)

Plants, flowering plants in particular, constitute the majority of the planet's biomass, and they are unquestionably the most visible element of life on Earth. And their existence, in turn, is inseparably tied to that of insects. Looking back at the evolutionary record of life on Earth, it becomes immediately apparent that the explosion of plant and insect diversity began at the same time, around 130 million years ago, in the Cretaceous Period. These two groups have been coevolving ever since, in a process that is at once a vicious arms race and relationship of mutual dependence. Indeed, over 200,000 species of flowering plants rely on insects for their reproduction. Pollination by insects has driven the origin of bromeliads, orchids, heliconias, passion fruits, gingers, and thousands of other iconic flowers for which Costa Rica is famous. Flowers of many plants can only be pollinated by certain species of insects, and their morphology precludes other insects from

Slug moth (*Perola producta*)

Hawk moth (*Callionima denticulata*)

Tiger moth (*Procalypta subcyanea*)

Silk moth (*Syssphinx quadrilineata*)

One of the most species-rich groups of insects is the moths and butterflies (order Lepidoptera). They have achieved their success thanks to a coevolutionary relationship with plants.

having access to nectar or other rewards. By visiting flowers, insects gain a wide range of benefits, including food (nectar, pollen, or both), convenient (and romantic) meeting sites for reproduction, and free heating (yes, some flowers generate heat that insects use to warm up their bodies). Not surprisingly, over time insects have evolved structures and behaviors that serve no other purpose but to access flowers. The long, extraordinarily modified mouthparts of butterflies and moths that perfectly fit the extended, nectar-filled corollas of flowers are a good example.

At the same time, leaves and stems of plants are busy producing spines and irritating hairs—and pumping their tissues with toxins—to repel or kill insects eager to feed on them. But for every new deadly compound that a plant evolves, a new mutation in an insect's genome will sooner or later make it possible

Flies (order Diptera) are the third most species-rich group of insects, surpassed only by beetles and wasps. While about 130,000 species have been described, at least four times as many await discovery. They can be found in virtually all terrestrial and aquatic habitats of the globe.

Leaf beetle (*Monocesta* sp.)

Beetles (order Coleoptera) compose not only the largest group of insects but the most species-rich group of organisms on Earth. An estimated 400,000 species have already been described by science, but many more remain to be discovered. Like moths and butterflies, the evolutionary success of beetles is closely tied to that of plants. The coevolutionary arms race between insects and plants has propelled both groups to great heights of morphological, physiological, and behavioral sophistication.

to overcome the new defense, which will then trigger a different response from the plant, and so on. The result of this back-and-forth process is an ever increasing rate of speciation in both plants and insects; new species appear, each with more sophisticated weaponry to attack or defend itself, and each with a weakness that will sooner or later be exploited. The symbiotic relationship between insects and plants is certainly one reason why insects have experienced such unparalleled success, but it is not the only one. Several key evolutionary innovations have helped them achieve dominance among all other animals—and maintain it for the last 300 million years. The first thing to notice about an insect's body is how small

it is, at least compared to our gargantuan frames. But small size confers many advantages. First, it makes it much easier to travel, either by active flight, by being carried by wind or water, or by hitching a ride on the body of a larger animal. This means that insects can colonize new areas and fill unoccupied niches more quickly. Smaller body size also makes it easier to hide from predators. Even more importantly, small bodied organisms require fewer resources to live; while a clump of grass is barely a snack for a cow, it can support hundreds of small insects for several generations. The one major disadvantage of a small body is that is loses water rapidly. Insects have dealt with this problem by evolving an

Rhinoceros beetle (*Coleosis biloba*)

impermeable exoskeleton made of chitin, a flexible and durable polysaccharide. The exoskeleton not only stops water loss but provides the animal with an exceptional degree of protection from injury. It also allows for the evolution of an almost inexhaustible variety of shapes and morphological defenses that increase the insects' ability to survive in the most inhospitable environments.

Compared to insects, human bodies seem to be built rather illogically; our soft skin barely covers vulnerable tissues and organs, while the hard, supportive skeleton is hidden within. With the exception of our brain, which is protected by a hard skull, everything else is pretty much exposed. A fall from a tall tree will result in massive damage to our muscles and organs, or may even be fatal. But drop a beetle from the tallest rainforest tree, and it will bounce and walk away unharmed. In insects all vulnerable elements of the body are safely encased within a rigid, impervious exoskeleton.

Insects were the first animals to evolve wings and powered flight, a feat repeated only three more times in the history of life (first by the now extinct pterodactyls, later by birds and bats). The ability to fly gave insects unprecedented advantages over other organisms by allowing them to disperse and colonize new areas and to escape predators. Although it certainly did not happen exactly in this way, one can imagine the surprise of the first predator whose prey flew up into the sky right from under its nose; it is as though the animal that you were hunting suddenly opened a portal to another dimension and vanished.

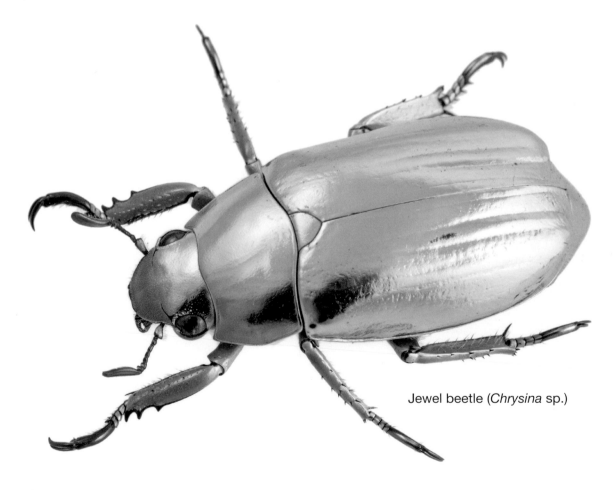

Jewel beetle (*Chrysina* sp.)

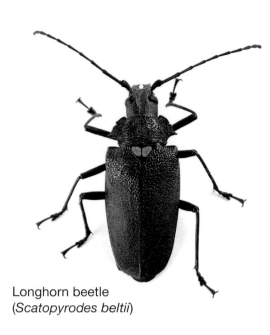

Longhorn beetle
(*Scatopyrodes beltii*)

Bess beetle
(*Passalus* sp.)

Straight-snouted weevil
(*Arrhenodes* sp.)

Harlequin beetle
(*Acrocinus longimanus*)

Top: Grasshopper (*Tropidacris cristata*). Bottom: Conehead katydid (*Copiphora cultricornis*).

How wings evolved is still a matter of discussion among biologists. The prevailing theory postulates that they are modified outgrowths of limbs, but recent fossil discoveries, combined with evidence from developmental biology and genetics, suggest a more complex mechanism. Apparently, chitinous plates on the dorsal surface of the thorax (tergites) co-opted genetic mechanisms that normally control jointed legs, eventually leading to the origin of movable wings. The first winged insects were undoubtedly rather clumsy fliers, probably similar in their morphology to today's mayflies or dragonflies. And the early wings could not be folded, which limited insects' ability to hide.

The next evolutionary innovation came with the ability of insects to fold the wings flat on the back, making the entire body more compact. That made it easier to squeeze into nooks and crannies, where they could hide, raise their young, or look for food. Blattodeans, an ancient and still thriving group of insects, are a good example of the benefit bestowed by this development. Once the wings could be folded, they could also start playing a more protective role, and in several lineages of insects the first pair of wings have turned into tough, sometimes rock-hard covers that shield the membranous hind wings and the soft abdomen. The tegmina of true bugs and the elytra of beetles are examples of such modifications.

Although the exoskeleton is doubtless a highly beneficial trait of insects, it is also their Achilles' heel. Because there is one thing that the exoskeleton cannot do—it cannot grow. Or at least it cannot grow gradually, slowly expanding to accommodate larger organs and a greater volume of body fluids. Instead, the entire exoskeleton has to be replaced at once by a brand new, larger version, which is what happens during the process of molting. In order to grow, each insect must periodically shed its old exoskeleton and rebuild a new one. In many insects, a molt also marks a change into a very different morphology and lifestyle.

Mantisfly (*Dicromantispa* sp.)

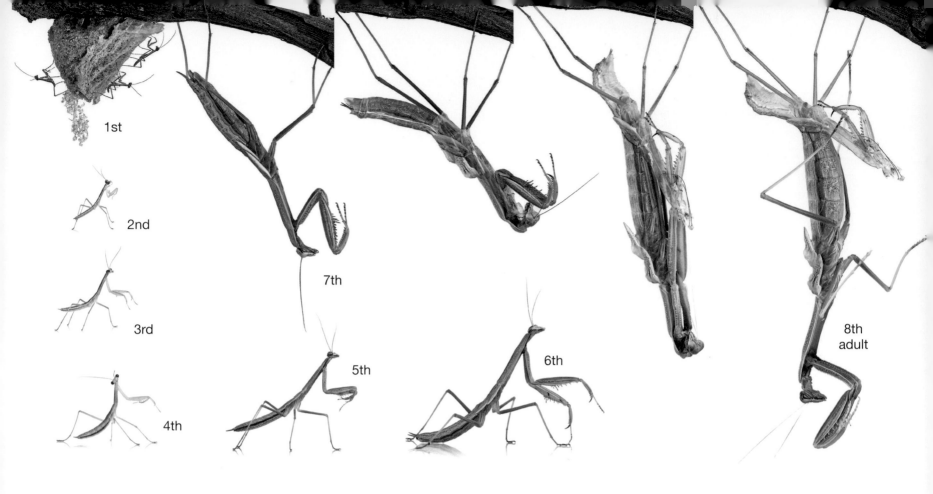

1st

2nd

3rd

4th

5th

6th

7th

8th
adult

In insects that undergo incomplete metamorphosis, the juvenile stages, or nymphs, are usually similar in their morphology and behavior to the adults. Each molt results in the gradual growth of the insect's body but only the last molt ends with the development of wings and reproductive organs. This praying mantis transitions from the last nymphal stage (7th) into the adult (8th) over several hours, during which time her wings unfold and begin to fill with air and hemolymph, eventually achieving their final appearance, which can be seen in the fully mature adult.

There is nothing in human biology and behavior that compares even remotely with the process of insect metamorphosis. Our birth, as dramatic an event as it is, merely represents a moment of emergence as a fully formed individual after a period of steady gestation in the womb, with virtually all organs and senses already in place. Human development is boringly gradual, with no instantaneous morphological transformations to punctuate the passage from childhood to adolescence, from adolescence to puberty; we do not grow or lose an extra set of appendages overnight, and we cannot replace our skeleton if it becomes too small. As children, we essentially eat the same food as when we are adults, and live in exactly the same environment. None of this is of course a bad thing—there is safety in this uneventful constancy. As beautiful and as fascinating as insect metamorphosis might be, it is also a moment of ultimate vulnerability; weak and immobile, a molting praying mantis can be killed by a small cricket, and a hawk moth emerging from its pupal case can only try to crawl away from danger, unable to fly as it waits for its powerful wings to

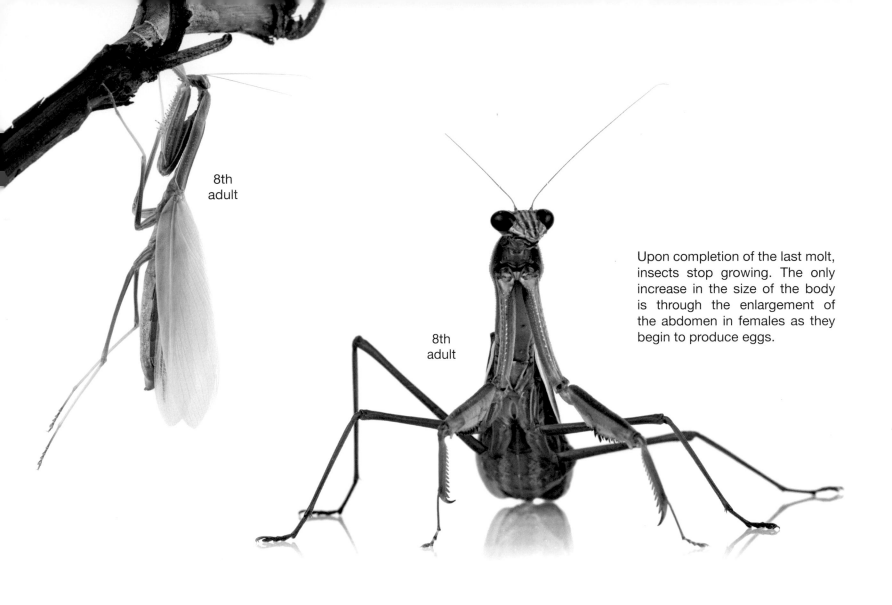

8th
adult

8th
adult

Upon completion of the last molt, insects stop growing. The only increase in the size of the body is through the enlargement of the abdomen in females as they begin to produce eggs.

dry and become functional. Every day untold trillions of insects undergo metamorphosis from a larva to a pupa, and from a chrysalis into a winged adult, and every day a large percentage of them die in the process.

The necessity for molting (ecdysis), a process that marks the transition from one stage of an insect's life into another, has to do with the structure of their skeleton. Unlike a vertebrate's calcareous internal skeleton, which provides muscle attachments and does not constrain the growth of tissues surrounding it, insects' chitinous exoskeleton has a limited ability to expand and must be shed periodically to accommodate growing tissue and newly appearing organs. The process of molting involves a complex sequence of cellular events, including a programmed cell death (apoptosis) that eliminates structures unnecessary during the next developmental stage. Molting is regulated by a series of hormones, primarily the prothoracicotropic hormone, which initiates the molt, and ecdysone, which controls the formation of new tissue.

What is even more interesting is the reason for the often dramatic differences between the adult and larval stages of insects, or even between larval stages of different ages. Entomologists classify insect developmental cycles into three broad categories, the simplest one being the ametabolous, found in primitive insects, such as silverfish. In these insects the young individuals differ from reproductive adults only in their size, and molting continues indefinitely, even after reaching sexual maturity. In hemimetabolous insects, or insects with direct development, immature forms have a body structure that resembles that of adults, but lack wings and reproductive organs, and usually lead a lifestyle similar to that of adults. Most insects with direct development have larvae (often referred to as nymphs) very similar to adults in their form and function—a young grasshopper looks like a grasshopper, even if it lacks wings and external hearing organs, and it usually feeds on the same things as its parents. A rare exception is the development in dragonflies, whose nymphs, known as naiads, lead an entirely aquatic lifestyle, very different from the adults. Such dramatic differences between larval and adult lifestyle are the rule in holometabolous insects, or insects with indirect development. In beetles, butterflies, and wasps the larva is so drastically different from the adult that it is often nearly impossible to identify them as members of the same species. The typical sequence of indirect development includes an egg, a mobile larva whose only goal in life is to eat and eat some more, an immobile and non-feeding pupa, and the reproductive adult (imago).

A cursory comparison of insects with direct (hemimetabolous) versus indirect (holometabolous) development immediately reveals that the latter have been far more successful:

pupa

larva

egg

imago

During complete metamorphosis, insects undergo a series of dramatic transformations and radical rearrangements of their external and internal morphology. The differences in the larval morphology and lifestyle means that the younger generation does not compete against their parents for the same resources.

The insect exoskeleton is one of the main reasons for the success of these animals, as it provides unparalleled protection against injury and water loss. At the same time, however, it limits insects' ability to grow as large as their vertebrate counterparts. Their growth is also limited by their respiratory system, which relies on largely passive gas exchange through a system of chitinous tubes, or tracheae. For these reasons few insects are as large as even the smallest of vertebrates. In Costa Rica, however, it is possible to witness true insect giants, such as this *Panthophthalmus* species, one of the largest flies in the world. These harmless giants feed on the roots of rainforest trees. Next to it sits a dart frog (*Oophaga pumilio*).

beetles alone, with over 400,000 described species, dwarf all hemimetabolous insects, and the evolutionary success of wasps, moths, and flies is similarly staggering. It appears that by finely partitioning their life cycle into periods of feeding (larva), major body restructuring (pupa), and reproduction (adult), during which process the immature stages do not compete directly for resources with the adults or each other, holometabolous insects have been able to dominate most of the world's terrestrial ecosystems. While a young praying mantis occupies the same niche and competes with its parents for the same food, an algae filtering mosquito larva might as well live on a different planet as far as its pollen- and blood-feeding parents are concerned.

In Costa Rica, a visitor will be able to witness insects' success in all their splendor. Few places in the world offer a similar opportunity to see such dramatic and beautiful examples of insects working as pollinators of rainforest plants such as the jewel-like euglossine bees and bright *Heliconius* butterflies. The work of leaf cutting ants will demonstrate the immensity of insects' contribution to the tasks of recycling organic material and of aeration and creation of the very soil. Spectacular Hercules beetles and rhinoceros katydids demonstrate the power of sexual selection in shaping the bodies and behaviors of animals. Hamadryas butterflies that make loud noises as they fly and moths that use ultrasonic cries to jam bats' echolocation will open a window onto the little known world of acoustic warfare as practiced by insects. The chapters in this book, while only an introduction to the Costa Rican fauna of insects and some of their arthropod cousins, which share many of insects' characteristics, will hopefully guide the reader on a fascinating journey of discovery of the most incredible members of the world's biodiversity.

Costa Rica is rich in unique, endemic species of insects, such as this Tico katydid (*Melanonotus tico*).

1

What Is This?

Insects are the most abundant and species-rich life form in terrestrial tropical habitats, and visitors to Costa Rica will soon encounter many examples of their incredible diversity. A lamp in front of a rainforest cabin, even a street light in the center of San Jose, will attract species strikingly different from those known to a visitor from North America or Europe, and a walk in the rainforest may lead to an encounter with a shockingly unfamiliar insect. Yet, while most Costa Rican species are different from those in temperate parts of the globe, they are all members of the same major lineages of insects, and share similar characteristics. This chapter introduces all orders of insects and a few related groups of arthropods that occur in Costa Rica, and highlights their most important features.

True insects, members of the class Insecta, are divided into approximately 30 orders, large groupings of related species united by their evolutionary history and often with similar appearance and biology. We use the word *approximately* advisedly, as some insect classification schemes divide a single order into several smaller ones, while others tend to lump smaller orders into larger groupings. It is thus helpful to keep in mind that orders, like all units of biological classification, are nothing more than artificial constructs that help us visualize the relatedness of the seemingly inexhaustible diversity of life. Their boundaries change as our knowledge of the evolution of life grows, helping reveal the often hidden and counterintuitive connections between seemingly dissimilar organisms.

Insects are members of the phylum Arthropoda, a larger group of animals that also includes arachnids (spiders, scorpions, and their kin), myriapods (millipedes and centipedes), and crustaceans (crabs, crayfish, and related animals). They are all characterized by having a rigid external skeleton (exoskeleton), composed of multiple plates and tubes that are connected by flexible joints. The exoskeleton is made of chitin, a strong and durable polysaccharide. In crustaceans and millipedes, the chitinous exoskeleton is reinforced with calcium, which makes it harder and more resistant to damage.

Older classification schemes included among the insects several lineages of small, six-legged arthropods that are now treated as separate, albeit closely related, classes. This chapter describes two such classes, the springtails and the diplurans. Included here are also several major classes and orders of non-insect arthropods that a visitor to Costa Rica is likely to encounter.

Despite displaying a dazzling array of shapes, colors, and sizes, insects are easy to tell apart from nearly all other invertebrate animals. First and foremost, if an invertebrate animal has wings, it is an insect—no other invertebrate is capable of active, sustained flight. And if the animal has three pairs of legs, a pair of large compound eyes, and one pair of antennae, it also cannot be anything else but an insect. There are some exceptions (nearly all orders of insects have wingless members), but in a great majority of cases these features will correctly identify an insect. Things get more complicated, however, once we try to identify larval

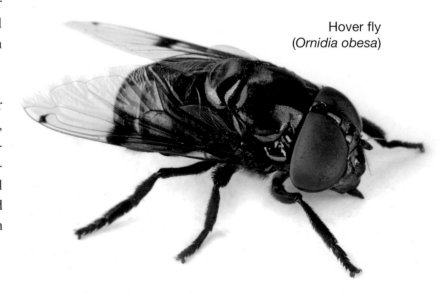

Hover fly
(*Ornidia obesa*)

Ironclad beetle (*Zopherus jansoni*)

stages of insects, which lack wings, sometimes lack eyes (or the entire head!), and may have many "legs" or rather leg-like structures (a caterpillar's multiple appendages, known as prolegs, are not true legs in the anatomical or developmental sense). Distinguishing among different orders of insects is also relatively easy. The brief descriptions below highlight the diagnostic characteristics of each order, but it is important to keep in mind that there exist scores of cases of morphological convergence among unrelated orders. Conversely, members of the same order may appear strikingly different from each other.

Before delving into the class Insecta, we begin with two related classes, the springtails and the diplurans.

Springtails (Class Collembola)

Although most species are tiny—less than a millimeter long—and usually overlooked, springtails are in fact some of the most abundant animals in the world. A cubic meter of soil may contain several hundred thousands of these organisms. Some species form dense mats of thousands or millions of individuals that can cover large areas of suitable habitat. Springtails can be identified by their ability to jump, a feat achieved with a spring-like structure (furcula) on the underside of their abdomen, and the presence of eyes composed of a small cluster of ommatidia (photoreceptor cells). These animals feed on algae and other plants, although some species hunt nematodes and other minute organisms. Springtails have little economic importance and are completely harmless to humans. Roughly 8,000 species are known worldwide, but only about 90 have been reported from Costa Rica.

A dense accumulation of springtails.

Diplurans (Class Diplura)

Once considered primitive wingless insects, diplurans are now placed in a separate class closely related to insects. Most are small (3–12 mm) and can be identified by the presence of a pair of distinct outgrowths (cerci) at the tip of the abdomen. This makes them similar to earwigs (order Dermaptera), from which they differ by the lack of eyes and wings. About 1,000 species are known, of which only 11 have been found in Costa Rica.

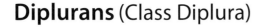

A springtail.

Dipluran (Japygidae)

Insects (Class Insecta)

Costa Rica's fauna includes all but three insect orders. Snakeflies (order Raphidioptera), members of an order closely related to lacewings (Neuroptera), occur in the northern hemisphere and only reach as far south as the mountains of Guatemala. Ice crawlers (order Grylloblattodea) are known from cold regions of North America and Asia. And heelwalkers (order Mantophasmatodea) only occur in arid regions of southern Africa.

Silverfish (Order Zygentoma) and Bristletails (Order Archaeognatha)

Silverfish and bristletails are true insects that nonetheless retain a number of ancient, simple features that characterized the ancestors of today's insects. They lack wings and continue molting throughout their lives, even long after reaching adulthood. Several species are found in houses, but they have no major economic impacts. About 1,100 species are known, most of them associated with arid habitats. Consequently, few have been found in Costa Rica.

Bristletail species.

Silverfish
(*Ctenolepisma lineata*)

Dragonflies and Damselflies (Order Odonata)

These easily recognizable aerial acrobats are some of the best fliers in the insect world. Dragonflies (suborder Anisoptera) are robust and have enormous eyes that envelop the entire head, giving them 360° vision. Damselflies (suborder Zygoptera) are more slender and have widely separated eyes. Adults in both groups feed on flying insects, including mosquitos and smaller species of their own order. Nymphs (known as naiads) of dragonflies and damselflies develop in water, and are predators of small fish, tadpoles, and aquatic invertebrates. About 290 species occur in Costa Rica, mostly at lower elevations in humid areas of the country.

Dragonfly (*Gynacantha tibiata*)

An unidentified mayfly.

Mayflies (Order Ephemeroptera)

Mayflies spend virtually all of their lives as aquatic nymphs, feeding on algae and detritus. During their development, nymphs may molt up to 50 times. After they emerge, the subimagos (winged but not fully matured) must undergo an additional molt before reaching the reproductive stage (imago). Mayflies are the only winged insects that molt after acquiring wings. The forewings are much larger than the hind wings, and are always held vertically over the body. Adult mayflies do not feed (indeed, they lack functional mouthparts) and only live for a few hours or days; their gut is filled with air to make their bodies lighter. Some mayfly species are very sensitive to pollution and low oxygen levels, making them important indicators of water quality. About 90 species of mayflies are known from Costa Rica.

Stoneflies (Order Plecoptera)

Associated mainly with fast-flowing, cold, well-oxygenated streams, stoneflies are most diverse in temperate areas of the globe; few species inhabit tropical regions. All stoneflies have a similar appearance, with an elongate body and wings folded flat on the back. After spending several months to a few years as aquatic nymphs, adult stoneflies emerge and immediately begin their courtship ritual. Stoneflies employ a sophisticated system of acoustic signals, composed of species-specific drumming patterns performed with the tip of their abdomen or the ventral side of the thorax. This drumming signal transmits itself through rocks in the water, allowing males and females to find each other in the noisy environment of fast-flowing streams. Like their close relatives mayflies, stoneflies are highly sensitive to water pollution and are used as environmental indicators of water quality. Only a handful of species, all in the genus *Anacroneuria*, are known from Costa Rica.

Stonefly (*Anacroneuria* sp.)

Grasshoppers, Katydids, and Crickets (Order Orthoptera)

Members of this order are easily recognizable thanks to their well-known ability to jump and produce sound. The hind legs of most orthopterans are saltatorial (leaping), with a large, muscular femur and a long, slender tibia. Some grasshoppers can perform repeated leaps of 2.6 m without any obvious signs of fatigue. This is possible because of the presence in their back legs of resilin, an elastic protein that has 97% efficiency in returning stored energy. This allows for explosive release of energy that catapults the insect, a task impossible with muscle power alone. Certain groups of orthopterans that lead a subterranean life, mole crickets, for example, have lost their ability to jump.

The three main groups of orthopterans—katydids, crickets, and grasshoppers—can be distinguished from each other by their antennae (short and thick in grasshoppers, long and thread-like in katydids and crickets), the position of the wings (held flat on the back in crickets, held vertically forming a "roof" in katydids and grasshoppers), and the shape of the egg-laying organ, or the ovipositor (laterally flattened in katydids, needle-like in crickets, and very short, nearly invisible, in grasshoppers).

Although the trait is less widespread than generally believed, many orthopterans are capable of producing loud sounds, though sometimes in frequencies that are difficult or impossible for the human ear to perceive. The role of sound production is threefold and similar in some respects to that of bird calls: attracting mates, announcing territoriality, and scaring off predators (alarm calls produced when seized by a predator). The calls of orthopterans are usually species-specific and play a very important role in species recognition.

The dominant mechanism of sound production in Orthoptera is stridulation, which involves rubbing one modified area of the body against another. Contrary to popular belief, no orthopterans produce sound by rubbing their hind legs against each other. Katydids and crickets produce sound by rubbing a modified vein (stridulatory vein) of one front wing (tegmen) against a hardened edge of the second one. Some katydids use their enlarged, shield-like pronotum (first thoracic segment) as a sound amplifier, whereas crickets often use other methods of sound amplification, such as singing from

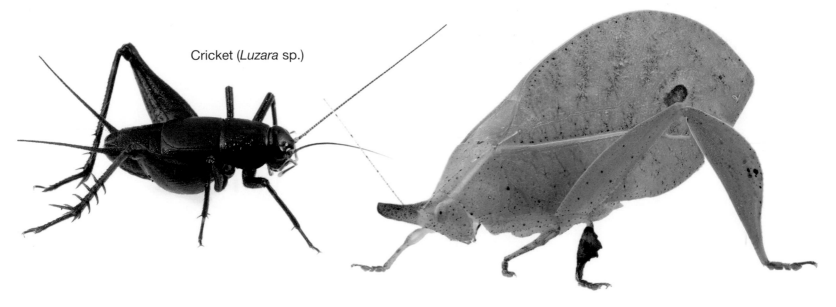

Cricket (*Luzara* sp.)

Katydid (*Aegimia maculifolia*)

burrows, the shape and size of which is attuned to boost certain frequencies (as in mole crickets), or using the surface of a leaf for the same purpose (tree crickets, for example). The ability to stridulate is restricted almost exclusively to males, although in some katydids females respond to the male's calls by producing short, often ultrasonic, clicks.

Grasshoppers use the same principle of stridulation, but instead of rubbing their wings against each other, these insects produce sound by rubbing the inner surface of the hind femur against one of the veins of the tegmen. In some grasshoppers (subfamily Gomphocerinae), the inner surface of the femur possesses a file of small knobs while the vein on the tegmen acts as the scraper. In others (subfamily Oedipodinae), the vein on the tegmen has a row of pegs and the femur plays the role of the scraper.

Orthopterans are extremely diverse in their food preferences and feeding techniques. Virtually all grasshoppers are strictly herbivorous, very rarely engaging in cannibalistic behavior, and doing so only under conditions of population crowding. Most grasshopper species are polyphagous (feeding on a wide variety of plant species), but some are oligophagous (feeding on a narrow spectrum of plant species) or monophagous (feeding on only one species of plant). The last are often associated with toxic, alkaloid-rich plant species, and these substances make the insects themselves inedible to many potential predators. Pygmy grasshoppers (Tetrigidae) are among the few insects that feed on mosses and lichens.

Katydids and crickets range from herbivorous to omnivorous to strictly predaceous. Most leaf katydids (Phaneropterinae) and sylvan katydids (Pseudophylinae) are strictly herbivorous, while the rhinoceros katydids (Copiphora), common in lowland areas of Costa Rica, feed on flowers, fruits, hard seeds, caterpillars, other katydids, snails, frog eggs, and even small lizards. Crickets tend to be generalists in their dietary preferences but rarely exhibit tendencies to feed on live prey. Some mole crickets have a unique behavior of gathering and storing germinating seeds in circular chambers below ground for later consumption.

Costa Rica has a rich variety of orthopterans, whose numbers in this country likely approach 1,000 species, some still unnamed by science.

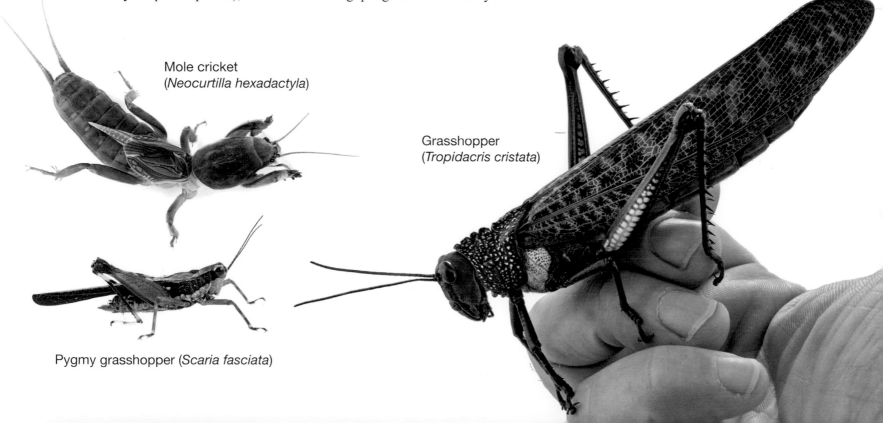

Mole cricket
(*Neocurtilla hexadactyla*)

Grasshopper
(*Tropidacris cristata*)

Pygmy grasshopper (*Scaria fasciata*)

Macromantis hyalina

Mantids, Blattodeans, and Termites

As counterintuitive as it may seem, praying mantids, blattodeans (cockroaches), and termites share a common ancestor and thus a large number of morphological and genetic features. So close is their relationship that some entomological classifications place these three groups of insects in a single order, Dictyoptera, though they are treated as separate orders for the purposes of this book. All three are characterized by direct (hemimetabolous) development, the presence of elongated coxae ("hips"), a large, shield-like pronotum, and a free-moving head with well-developed eyes (except for some castes of termites). In their behavior, however, these groups cannot be more different. Mantids are solitary predators; blattodeans feed on detritus and can be solitary or gregarious; termites, who feed on wood, are truly social insects with a sophisticated division of labor. As a group, the superorder Dictyoptera is one of the best examples of how forces of natural selection can lead to diametrically opposed lifestyles among closely related organisms.

Praying Mantids (Order Mantodea)

Their nearly human-like appearance and ability to turn their head to meet a person's gaze, combined with a reputation for ferocity and cannibalism, make praying mantids stand out among all the insects. Their most conspicuous characteristics are the large, raptorial front legs; long, neck-like first segment of the thorax (prothorax); and a triangular head with large, widely separated eyes. Most species of praying mantid are sit-and-wait predators, usually perfectly camouflaged against their background. They feed mostly on insects, although larger species are capable of capturing small vertebrates such as lizards and hummingbirds. An unusual feature of praying mantids is that they are equipped with a single ear, located between the coxae of their hind legs. This means that these insects lack the ability of directional hearing; even so, their single ear can detect the ultrasonic signals of echolocating bats, helping them avoid these predatory mammals. Although some species do exhibit sexual cannibalism, this behavior is not very common. Female mantids lay eggs in large, protective cases (ootheca), and in some species guard the eggs against predators and parasitoids. There are 63 species recorded from Costa Rica, with more expected to be found.

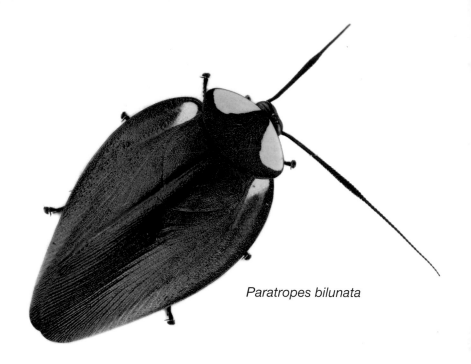

Paratropes bilunata

Blattodeans (Order Blattodea)

These fascinating insects exhibit complex maternal behaviors that rival those found in mammals and birds. Some species encase eggs in a protective case (ootheca); others carry eggs until the moment of hatching; and yet others give birth to live young that the mother carries on her back until they are ready to fend for themselves. In some blattodeans, the female develops the equivalent of a mammalian placenta, while in a few species the mother suckles her young with special "mammary glands" between the coxae of her legs. Blattodeans are abundant in the leaf litter of the rainforest, where they play an important role recycling dead organic matter; a few species act as pollinators of rainforest plants. Some diurnal species exhibit beautiful coloration, while others mimic in their appearance fast moving wasps and flies, in all likelihood to discourage potential predators from pursuit. *Megaloblatta blaberoides* is one of Costa Rica's largest insects. Nymphs of this species are capable of producing sounds that resemble that of a rattlesnake's rattle, presumably to warn predators of sticky, repellent chemical compounds present in their abdomen. The fauna of Costa Rican blattodeans is poorly known but over 100 species are likely to be found here.

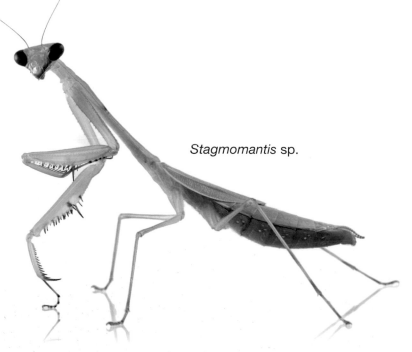

Stagmomantis sp.

Rhynchotermes perarmatus

Termites (Order Isoptera)

Termites are one of only a handful of animal groups that have evolved eusociality, or true social behavior, in which different members of the society perform distinct, specialized tasks. In eusocial societies only select individuals can reproduce, while sterile individuals are given the role of protecting the reproductive individuals and maintaining the cohesion of the colony (this means that the human society is not fully eusocial). A typical termite colony includes the castes of workers, soldiers, the king, and the queen. Unlike ant societies, where all non-reproductive members are sterile females, both males and females can function as workers and soldiers. The king and queen of the termite colony are remarkably long-lived: there is evidence that in some termite species the queen may reach the age of 70, making her the longest living insect.

Termites feed on plant material, including wood, which they are able to digest thanks to the presence of symbiotic protozoans in their gut as well as the ability to produce enzymes that digest cellulose. Some highly derived lineages of termites practice agriculture by growing fungi in their nest. In Costa Rica, 45 species of termite have been recorded but undoubtedly more await discovery.

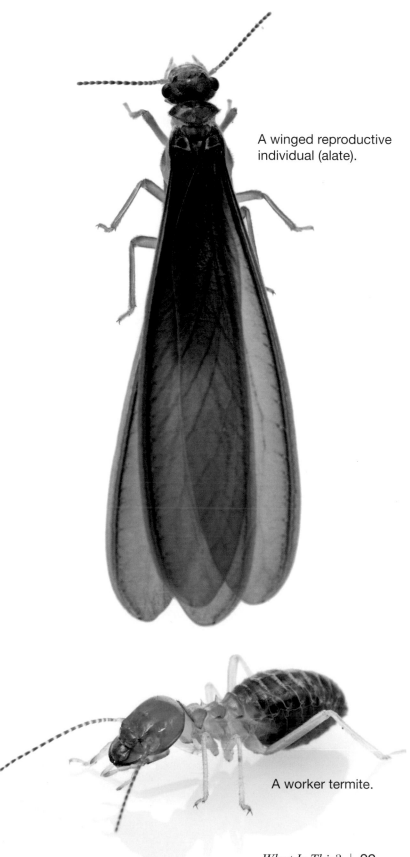

A winged reproductive individual (alate).

A worker termite.

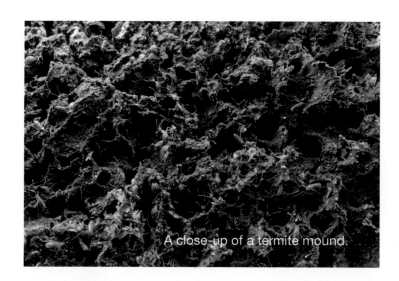

A close-up of a termite mound.

Phanocles sp.

Trychopeplus laciniatus

Rhynchacris bigibbus

Stick Insects (Order Phasmida)

With a body length exceeding 20 cm, some stick insects are the longest insects to be found in Costa Rica, although their incredible resemblance to plants makes them virtually invisible. Unless they are spotted when moving, usually at night when they feed on leaves, a visitor to the rainforest is unlikely to notice one. Even stick-insect eggs, which are perfect replicas of seeds, are difficult to find. In addition to having amazing camouflage, some species of stick insect are capable of synthesizing toxic compounds that make their bodies unpalatable to predators. Such species usually have bright markings on their body to warn potential predators of their toxicity. Stick insects are frequently sexually dimorphic, which means that males and females may look very different. In some species, males possess fully developed wings while the females are wingless. The fauna of Costa Rican stick insects is poorly known, but about 100 species are likely to occur here.

Phanocles sp.

Earwigs (Order Dermaptera)

Recognizable by their strong "forceps" (cerci), and feared for their presumed proclivity to crawl into people's ears, earwigs are completely harmless and not more likely to end up in a person's ear than any other insect. Their cerci have a number of uses, one of which is helping fold their large, membranous hind wings underneath their tiny, hard forewings (tegmina). The cerci are also used in defense, in courtship, and for capturing prey (smaller insects). Earwigs display well-developed maternal behavior; the female stays with the clutch of eggs, cleaning and protecting them until her children hatch. There are about 100 species of earwig known from Costa Rica.

Metrasura flaviceps

Ancistrogaster sp.

Zorapterans (Order Zoraptera)

These tiny insects are rarely noticed, as they spend their lives inside rotting logs, where they feed on fungi and small invertebrates. Despite their inconspicuous appearance, zorapterans have a fascinating biology, with older males maintaining large harems that they defend from younger males. Their reproduction is also unusual, as the males produce only a single sperm cell, which is as long as the entire body of the insect. Zorapterans are one of the smallest orders of insect (39 species worldwide) and only a few species are known from Costa Rica.

Zorotypus neotropicus

Webspinners (Order Embioptera)

Webspinners are small, usually wingless, insects that are rarely noticed by the casual observer. Their colonies, in the form of sheets or strips of white silk on tree trunks, are likely to be mistaken for the work of spiders. Webspinners produce silk from special glands on their front tarsi; they use the silk to build long corridors and shelters over lichens (on which they feed) growing on tree bark or rocks. These insects are semi-social and often form aggregations of individuals of different ages. Only six species of webspinner have been recorded from Costa Rica but their numbers are undoubtedly higher.

Webspinners under a sheet of protective silk.

An unidentified webspinner (note enlarged tarsi on the front legs).

True Bugs, Cicadas, and Relatives (Orders Heteroptera, Auchenorrhyncha, and Sternorrhyncha)

These three groups of closely related insects have historically been treated as a single order, Hemiptera, but recent phylogenetic work supports the idea that dividing them into distinct orders better reflects their evolutionary history. Although they show an incredible amount of morphological diversity, all three groups have mouthparts that have been modified to form a needle-like sucking organ. This means that these insects can only ingest food in liquid form. True bugs (order Heteroptera)—named so as to distinguish them from other insects given the colloquial term *bug*—include herbivorous, predacious, and parasitic species, whereas members of the remaining two orders feed exclusively on plant sap. Assassin bugs (members of the family Reduviidae) and other species that feed on other insects often inject their victims with venom and digestive enzymes to dissolve all soft tissue. Some assassin bugs and other predaceous species may deliver a very painful bite if handled. Most true bugs are equipped with glands that produce compounds that smell and taste bad, which makes them unpalatable to many predators. Stink bugs (family

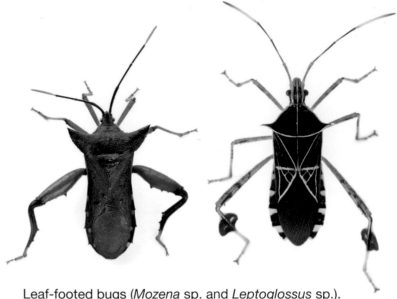

Leaf-footed bugs (*Mozena* sp. and *Leptoglossus* sp.).

Pentatomidae) are known for this behavior, but many other species have equally well-developed chemical defenses, and such species display bright, often quite beautiful, warning coloration.

True bugs can distinguished from the two remaining, closely related, orders by the structure of their front wings, each of which is divided into a hardened, opaque basal half and a membranous, often semi-translucent, apical half. Shield bugs (family Scutelleridae) have their wings covered entirely by a hard, convex shield (scutellum), making them look very beetle-like; they can be distinguished from beetles by their needle-like mouthparts (beetles have biting mouthparts with a pair of mandibles).

Some true bugs parasitize larger animals by feeding on their blood. The best example of such behavior is the infamous bed bug (*Cimex lactularius*), a cosmopolitan pest that has in recent years achieved international notoriety by becoming more common and by developing resistance to many pesticides used to combat it.

Members of the order Auchenorrhyncha include cicadas (family Cicadidae), which are famous for their ability to produce very loud if not melodious calls (you can read more

Assassin bug (*Zelurus* sp.)

about how they produce their calls in Chapter 5). Cicadas spend their early lives as subterranean, wingless nymphs equipped with large, shovel-like front legs that allow them to move underground in search of the plant roots on which they feed. Their time for development may be extremely long. The North American periodical cicadas can take 17 years to develop, and some tropical species may also take many years to reach maturity. Adult cicadas feed on xylem sap, watery liquid of low nutritional value. In order to extract enough nutrients from it they must imbibe massive quantities of the liquid, excreting the excess water by ejecting it from the anus. Standing under a tree full of feeding cicadas often feels like standing in a light rain.

Other families of Auchenorrhyncha feed on phloem sap, which contains much higher concentrations of sugar. They also excrete large quantities of sugary water, known as honeydew, which is eagerly sought out by many animals, most notably ants. Treehoppers (family Membracidae) and other related insects are often actively defended by ants from wasps and other predators in exchange for droplets of honeydew. Treehoppers not only tolerate the ants but often respond to their requests for more honeydew, which the ants indicate by tapping the treehopper's abdomen with their antennae.

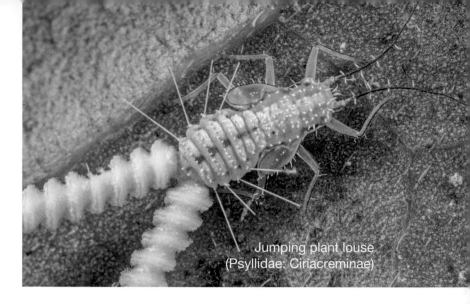
Jumping plant louse
(Psyllidae: Ciriacreminae)

The third suborder, Sternorrhyncha, includes very small, plant-feeding insects such as aphids, mealybugs, and scale insects. So adapted to a sedentary life on plants, scale insects have lost most of their appendages, including the legs. Many species in this group of insect produce large amounts of white wax, sometimes in the form of long threads at the tip of the abdomen. The function of the wax is not entirely understood, but it likely helps the insects evade predators, who may grab the breakable, easy-to-replace structure. Waxy coating also likely helps retain water in these sedentary species.

All three orders of hemipterans are very species rich, and it is likely that Costa Rica has several thousand species of these insects.

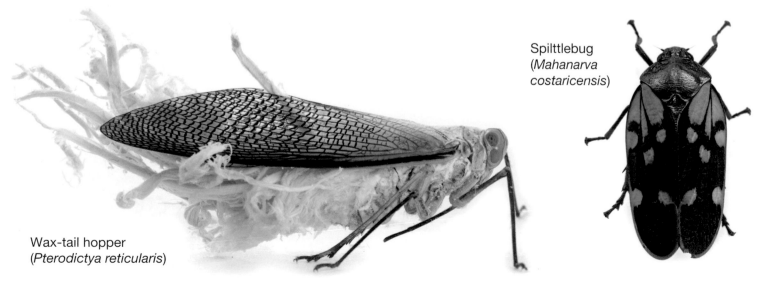

Wax-tail hopper
(*Pterodictya reticularis*)

Spilttlebug
(*Mahanarva costaricensis*)

Barklouse (*Poecilopsocus* sp.)

Barklice (Order Psocoptera)

Often only about a few millimeters long, barklice are tiny insects that sometimes form large aggregations on tree trunks or leaves. A few species enter houses, where they feed on mold growing on paper, earning them their common name, booklice. The species living in Costa Rica feed primarily on algae and fungal spores. A few large species display bright coloration and at first glance might be confused with small wasps.

Lice (Order Phthiraptera)

These exclusively parasitic insects are associated with a number of mammalian and avian hosts. Lice feed on either blood (sucking lice) or skin and feathers (chewing lice), and many are highly host-specific. Some are vectors of diseases; the human louse, for example, can transmit typhus. Most species, however, pose no danger to people, and a few are endangered because of the threatened status of their hosts. The exact number of lice species in Costa Rica is unknown but over 50 species have already been recorded.

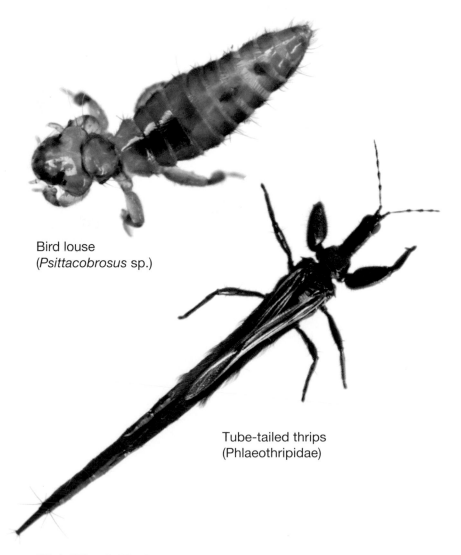

Bird louse
(*Psittacobrosus* sp.)

Tube-tailed thrips
(Phlaeothripidae)

Thrips (Order Thysanoptera)

Only a few millimeters long, these small insects have a number of characteristics that set them apart from all other member of the class Insecta. They can be identified by their unusual wings, which are fringed with long "hair" (setae), and the presence of strongly asymmetrical mouthparts that possess only one mandible (on the left), modified into a sucking organ. Thrips occupy an intermediate position between insects that undergo direct metamorphosis and those that undergo indirect metamorphosis: their larval stage is followed by up to three, non-feeding pupal stages; in some species the pupae spin a silken cocoon. The Costa Rican fauna includes about 160 recorded species, although more are expected to be found.

Hangingflies (Order Mecoptera)

Hangingflies are predaceous insects that are uniform in their appearance and resemble crane flies. Crane flies, which are true flies (order Diptera), have only one pair of wings (two pairs hangingflies). These insects are sit-and-wait predators of moths, flies, and other small insects, which they catch with their outstretched middle and hind legs while hanging by their front legs from vegetation (hence their common name). The tarsi (feet) of hangingflies are grasping, capable of closing like a hand, but the price for this ability is that that these insects are incapable of walking and can thus only hang or fly. Hangingflies have a complex courtship behavior, in which males offer females nuptial gifts in the form of freshly caught insects. Only 7 species of hangingfly are known from Costa Rica.

Fleas (Order Siphonaptera)

All members of this insect order are external parasites of warm-blooded vertebrates. They are distantly related to hangingflies, although their lifestyle could not be more different. Several species of flea are vectors of diseases, including the bubonic plague. The chigoe flea (*Tunga penetrans*) has the unpleasant habit of burrowing into the feet of people walking barefoot in the sand, which may lead to serious, even fatal, tetanus infections. With the exception of a handful of common species associated with domestic animals, the Costa Rican flea fauna is poorly known.

Hangingfly
(*Bittacus* sp.)

An unidentified flea.

Antlion larva
(*Myrmeleontidae*)

Lacewing
(*Nallachius americanus*)

Mantisfly
(*Climaciella brunnea*)

Antlions, Lacewings, and Relatives (Order Neuroptera)

Members of the order Neuroptera are considered the oldest lineage of insects that undergo indirect development. All species are predaceous, at least in their larval stage. Neuropterans have two pairs of large wings of similar size and venation, and some species (antlions and owlflies) may be mistaken for dragonflies, from which they differ in having thick antennae, often with clubbed tips.

The best known members of the Neuroptera are the antlions (family Myrmeleontidae), whose conical trap pits dot many sandy areas in Costa Rica. At the bottom of each pit, buried deep in the sand, sits a single antlion larva. Any small insect unlucky enough to stumble into the pit is met with a barrage of pebbles catapulted by the antlion larva, with the pebbles sending the victim tumbling towards the sharp, wide open mandibles. Interestingly, antlion larvae do not have a mouth opening with which to consume the victim, and instead pierce the prey with their mandibles, inject it with digestive enzymes, and then suck the content of its body dry.

Mantisflies (family Mantispidae) are neuropterans that resemble small praying mantids, from which they can be distinguished by the vertical way in which they hold their raptorial front legs (praying mantids hold their raptorial legs horizontally). Larvae of mantisflies are specialized predators of spider eggs and develop inside the spider egg sacks, often right under

Green lacewing
(*Chrysopidae*)

the watchful eye of the female spider, who is unaware of the demise of her brood. Eighty-five species of Neuroptera have been recorded from Costa Rica.

Dobsonflies (Order Megaloptera)

Dobsonflies are large, ferocious-looking insects that are often equipped with sharp mandibles. Despite their appearance, they are completely harmless and as adults feed only on nectar or fruit juices. Members of the genus *Corydalus* are very sexually dimorphic, with the males equipped with a pair of tusk-like mandibles. They use them in ritualized male-to-male combat, but because the mandibles have scant musculature, they cannot use them as true defensive weapons. In the genus *Platyneuromus*, males have on their heads large, wide plates whose function is unknown. Larvae of dobsonflies are aquatic and prefer well-oxygenated streams, although some species can develop in tree holes filled with water that contains low oxygen levels. Dobsonfly larvae are very familiar to recreational fishermen as hellgrammites and used as bait. Costa Rica has only a few species of dobsonfly, assigned to 3 genera.

Dobsonfly
(*Platyneuromus soror*)

Corydalus sp.

Chloronia sp.

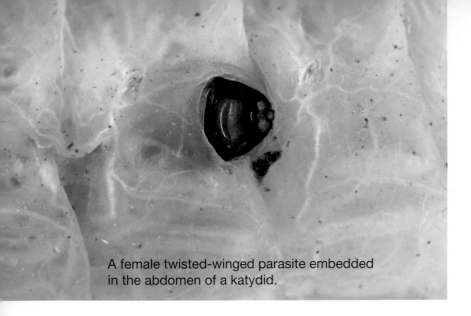

A female twisted-winged parasite embedded in the abdomen of a katydid.

Twisted-winged Parasites (Order Strepsiptera)

These highly specialized internal parasites of insects are sometimes noticed as small protrusions on the abdomen of bees, katydids, and other insects. A dark plate sticking out from between the abdominal segments is the head of the parasite; the remainder of the parasite is an amorphous sack permanently embedded in the tissues of its host. Males, which often develop in different hosts than the females, are free living and use their single pair of wings to search for females. Costa Rican species are poorly known, and only a handful have been found, on katydids and ants.

Beetles (Order Coleoptera)

With over 400,000 known species, and perhaps several times as many undiscovered ones, beetles are the largest, most species-rich, lineage of organisms on the planet. Their diversity is strongly linked with that of flowering plants, with which they have had an ancient coevolutionary relationship dating back to the Jurassic. Beetles live in all terrestrial habitats and most aquatic habitats—they do not occur in deep oceanic waters.

Beetle bodies are relatively uniform across most lineages of the order, displaying only superficial differences in proportion, shape, and color pattern. With few exceptions, the first pair of beetle wings is modified into hard shields known as elytra that protect a large, membranous, translucent hind wing that folds transversely underneath. In most groups the elytra also protect the two posterior segments of the thorax and the soft, upper side of the abdomen, although in groups such as rove beetles (Staphylinidae) and some longhorn beetles (Cerambycidae) the elytra are shortened and the abdomen is fully exposed. The first segment of the thorax (prothorax) forms a distinct, usually freely articulated (that is, not fused with other segments), body part. The antennae of beetles, regardless of their length, usually consist of fewer than 11 segments. The mouthparts are constructed to allow for chewing; beetles have well-developed mandibles that in some species are greatly enlarged to serve as formidable weapons used for hunting (e.g., tiger beetles) or male-to-male combat (e.g., stag beetles).

Beetles undergo indirect metamorphosis; larval development can be quite long, and large species may take several years to reach adulthood. A few groups of beetles that are parasitic on other insects undergo hypermetamorphosis, a type of development with an additional, free-living larval stage known as the planidium. Beetle larvae usually occupy niches that are very different from those of the adults.

Leaf beetle (*Diabrotica regalis*)

Weevil (*Eurhinus magnificus*)

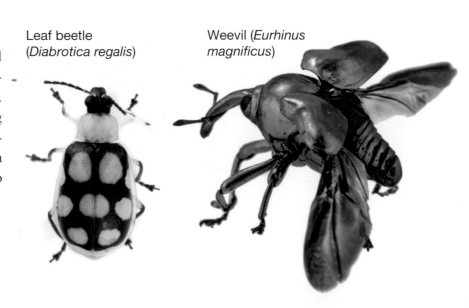

Heterosternus oberthuri

The beetle diet includes nearly every imaginable substance of organic origin, from live and decomposing plants and fungi, to wood and sap, to live insects and small vertebrates (there arc beetles whose larvae specialize on hunting frogs), to the carcasses of animals (including bone and horn). While few adult beetles are parasitic, a lineage of aquatic beetles parasitizes beavers and otters, and larvae of many beetles are parasitic on insects. In the rainforest ecosystem of Costa Rica, beetles play roles of herbivores, wood borers, and decomposers of dead organic matter. They pollinate plants and disperse fungal spores. Many are hunters, while others are food for birds and other vertebrates. Social insects such as ants and bees host many beetle species within their colonies, where the beetles may simply steal their food or feed on their hosts.

The beetles of Costa Rica have not been comprehensively studied, but the country certainly includes tens of thousands of species, many still awaiting formal description.

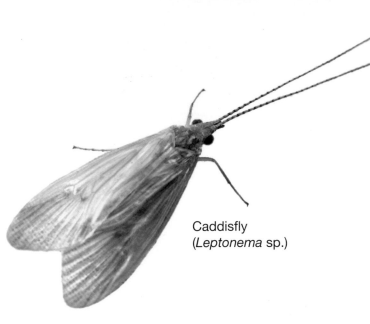

Caddisfly
(*Leptonema* sp.)

Caddisflies (Order Trichoptera)

Closely related to butterflies and moths, caddisflies spend most of their lives as aquatic larvae, often in intricately constructed silken shelters that are sometimes reinforced with grains of sand or small twigs. Many species are filter feeders, collecting organic material and small organisms suspended in the water column with the help of a net made of silk or long hairs on their legs. Others collect bits of organic matter from the bottom of streams and lakes, while others are predators of small aquatic invertebrates, including sponges.

Adult caddisflies are usually short-lived and feed only on liquids, such as flower nectar. Their wings are held roof-like over the body and are covered with dense hair or scales, and some resemble moths, from which they can be distinguished by the lack of a spirally curled proboscis. About 500 species of caddisflies are known from Costa Rica.

Moths and Butterflies (Order Lepidoptera)

Members of this order are undoubtedly some of the most recognizable insects—and among the most beautiful. Their two pairs of wings covered with scales that form intricate color patterns and mouthparts modified into a coiled, tubular proboscis are unique among all insects. There are exceptions to this rule, as both wingless moths as well as moths with reduced mouthparts exist, but in general moths and butterflies can be identified at a glance.

Lepidoptera is one of the largest orders of insects, with over 160,000 species already named and at least as many awaiting discovery and description. One key to their success—and diversity—is a combination of unique morphological traits. Anybody who has ever tried to hold a moth between two fingers quickly realizes that these insects are incredibly slippery. The microscopic scales that cover every surface of a lepidopteran body are designed to rub off at the lightest touch, which allows them to escape from the beaks of birds and the grasps of other predators. The same scales also provide an almost limitless palette of colors that some species rely on to indicate their toxicity to potential predators. Colors are also used to advertise suitability as mating partners or to create camouflage that allows many species to seamlessly blend in with their environment. The defensive nature of the slippery scales is often augmented by the presence of long tails on the hind wings that act as flight stabilizers and also easily break off in the mouths of aerial predators, allowing the insect to escape relatively unharmed.

Moths and butterflies undergo indirect metamorphosis. Their larvae (caterpillars) lead a lifestyle that is entirely different from that of the parents. In essence, caterpillars are sedentary eating machines; between two consecutive molts, they are capable of increasing their body mass 15 times by consuming huge amounts of plant tissue. Their parents feed on nectar and pollen, or hardly feed at all, which means that they do not compete with their own offspring for the same resources.

The diversity of moths and butterflies is strongly linked to that of flowering plants, and is the result of an ongoing arms race between these two major groups of organisms. Plants continually evolve new chemical and mechanical defenses against voracious caterpillars, to which the lepidopterans respond by selecting genetic variants resistant to the defenses. This leads to ever increasing specialization and speciation (origin of new species) in both groups. The two groups are linked in yet another way, as moths and butterflies are the primary pollinators of plants. Numerous plants have evolved adaptations to attract only certain lepidopteran species. Costa Rica has a huge number of moths and butterflies, with at least 14,000 species already reported from the country. Many lineages of the order, however, are virtually unstudied and many more species are expected to be discovered.

The Colors of Butterflies

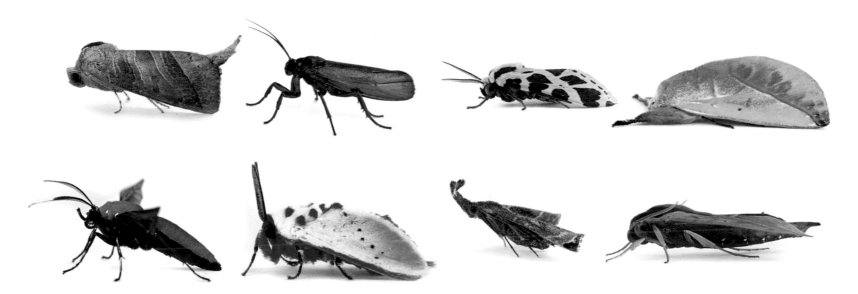

The astounding variety of lepidopteran wing patterns is the result of one of two mechanisms, or both: pigment-based coloration and structural coloration. Species with pigment-based patterns either synthesize their own pigments—including melanins, ommochromes, and porphyrins—or obtain them from plants (flavonoids and carotenoids). These pigments create different colors by selective absorption of particular wavelengths of the light spectrum. Absorbed wavelengths are subtracted from the total spectrum, while the remaining wavelengths are reflected to produce the visible color. Pigments range from dark and dull to extremely bright, but never show a metallic iridescence.

Species that employ structural colors do not rely on chemical compounds but rather on the microstructure of individual scales that produce optical effects through diffraction, refraction, and interference. For example, the metallic blue coloration of the iconic morpho butterfly is caused by the presence of tiny slits arranged 200 nanometers apart (a nanometer is one millionth of a millimeter) on scales of the wing. Because the blue part of the light spectrum has the wavelength of 400–480 nm, those wavelengths do not pass through the 200 nm slits but rather scatter in all directions, including towards the eyes of the observer, thus producing the transcendent color known as Tyndall blue.

Paper wasps (Polistinae)

Parasitic wasp (*Brachymeria* sp.)

Peleoinus polyturator

Wasps, Ants, and Bees
(Order Hymenoptera)

The fantastically diverse order Hymenoptera, which is estimated to include at least 300,000 species, is difficult to characterize, even though most of its common members—wasps, bees, and ants—are familiar to the general reader. The diversity in morphology and behavior in this order is unmatched by any other group of insects. There are few morphological features that unite all its members. Those species that have wings can be identified by the presence of a set of tiny hooks (hamuli) on the front edge of the hind wings that interlock with the posterior edge of the front ones. Thus, despite having two pairs of wings, functionally these insects have only one pair, as the front and the hind wings always move in unison. However, many groups of Hymenoptera are wingless, among them the ants, which are simply highly modified social wasps that evolved from winged wasp ancestors about 100 million years ago. Another characteristic present in members of the suborder Apocrita (to which most commonly encountered species belong) is the presence of a "waist," a distinctly narrowed section between the first and second segments of the abdomen. The waist allows for greater mobility of the abdomen, which gives wasps an advantage during oviposition (egg laying) and also better control of the stinger, the unique weapon present in many Hymenoptera. No other insect can use its ovipositor, the egg laying organ, to deliver venom, be it for defense or to immobilize its prey. For this reason, only female wasps and bees can sting, although males of many species will mock "sting" when threatened, without actually causing any harm. But psychological preconditioning will certainly make one brush off any wasp or bee that makes threatening movements.

Most Hymenoptera have chewing mouthparts with well-developed mandibles, although nectar feeding species such the

Velvet ant (Multilidae:
Sphaeropthalminae)

Zombie wasp
(*Ampulex* sp.)

honey bee (*Apis mellifera*) possess a highly enlarged lower lip (labium) that acts as a tongue for scooping up liquids.

All Hymenoptera share an interesting mechanism for determining sex. In most animals, the sex of an individual is dictated by a pair of chromosomes; in mammals, for example, females have sex chromosomes XX, whereas males have chromosomes XY. In the Hymenoptera, however, sex is determined by whether the egg from which the insect emerges has been inseminated or not. A fertilized egg always produces a female, an unfertilized egg produces a male. Several lineages of the Hymenoptera form complex societies, which may be related to this genetic mechanism by favoring cooperation among female workers, who are very closely related to each other (but note that some recent theories question this assumption). This behavior has reached its pinnacle in ants, whose societies are the most sophisticated among all animals, human societies included, in the degree of their division of labor and the ability to control their environment (you can learn more about this aspect of ant behavior in Chapter 7).

A large percentage of wasps are parasitoids of other insects (parasitoids always kill their host whereas parasites usually do not). Such species tend to have a very long ovipositor that allows them to penetrate the substrate, wood or other plant tissue where their victims, larvae of beetles and other insects, live. Amazingly, certain wasp species are hyperparasites—they lay eggs inside larvae of parasitoid wasps that develop in the larvae of other insects.

Hymenoptera are critically important members of many ecosystems, as pollinators, recyclers of organic matter, predators, parasitoids, and as the prey of other animals. In Costa Rican rainforests, the biomass of ants is approximately equal to that of all vertebrate animals present there. The close coevolutionary relationship between hymenopterans and plants means that many plants simply would not be able to exist without these insects. For example, all figs, common rainforest trees, are entirely dependent on tiny wasps that pollinate their flowers, for without them these plants would not be able to reproduce. Costa Rica has at least 20,000 species of Hymenoptera and likely many more await discovery.

Flies (Order Diptera)

Flies are annoyingly familiar to everybody, and many of the 150,000 known members of this order are external parasites that are often vectors of dangerous diseases. At the same time, the great majority of fly species are not only perfectly harmless to humans but provide invaluable ecological services without which many ecosystems would not be able to function. Flies can be identified by the unique structure of their wings. The front wings are large and membranous, and folded flat on the back when the insect is at rest. The hind wings are modified into tiny, club-shaped structures called halteres that act as gyroscopes, extremely sensitive balance organs that allow flies to perform aerial maneuvers impossible for other insects. Hover flies (family Syrphidae) and a few other fly families can maintain a virtually constant position midair while feeding on flower nectar, a feat that can only be matched, barely, by hummingbirds and hawk moths.

Some flies, especially those that lead a parasitic lifestyle as adults, have lost their front wings, but even these species retain their halteres.

The mouthparts of flies are modified for ingesting liquid food. In parasitic species such as mosquitos, the mouthparts form a long piercing apparatus, a tubular stylet (highly elongated mandibles and maxillae) encased in a protective sheet (a modified labium). Similar piercing structures are also found in predatory species such as robber flies (family Asiilidae), which hunt insects and kill them by injecting neurotoxic saliva and proteolytic enzymes that digest the victim's tissues. In species that feed on fruit, dung, or other organic matter, the mouthparts lap liquids with the notably widened, pad-like tips of the labial palps. These act like a piece of sponge, collecting liquids that are then transferred through a tubular labrum.

Fly larvae are highly diverse in their habitat preferences and morphology, but can be distinguished from those of other insects by the absence of jointed thoracic legs. Some fly larvae possess soft thoracic and abdominal prolegs, similar to those found in caterpillars, while others—known to entomologists by the charming name of creeping welts—have sets of movable, swollen areas equipped with setae ("hair"). Aquatic fly larvae such as those of mosquitoes or hover flies are good swimmers, and breathe using a long respiratory tube at the end of the abdomen.

Despite the deleterious impact of parasitic and disease-transmitting species on human health and the damage caused by pest species to agriculture, flies are one of the most critical elements of most natural ecosystems. A number of families (e.g., Calliphoridae, Sarcophagidae, and Muscidae) are decomposers and recyclers of organic matter, including carrion and dung. Aquatic larvae of many families, including midges (Chironomidae) and mosquitoes (Culicidae), are the principal food of fish and other aquatic organisms, whereas adults of the same flies form the basis of the diet of birds and bats. Hover (Sirphidae) and bee flies (Bombyllidae) are principal pollinators of many plants, as are mosquitoes (only the females of these insects feed on blood, whereas males feed exclusively on nectar). Costa Rica has an estimated 30,000 species of flies. In some groups such as aquatic Chironomidae, hundreds of species await formal description.

Deer fly (*Catachlorops fulmineus*)

Arachnids (Class Arachnida)

Arachnids are ancient animals whose ancestors inhabited Paleozoic oceans before venturing onto land; their close relatives, horseshoe crabs and sea spiders (Pycnogonida), are still strictly marine. Unlike insects, arachnids do not have a distinct head, which is fused with the thorax. They also lack antennae and wings. Most arachnids have 4 pairs of walking legs; some species (e.g., scorpions and pseudoscorpions) have large, pincerlike pedipalps, which are modified mouthparts. Arachnids can only ingest liquid or semi-liquid food, and thus many of them inject their food, be it plant material or an animal victim, with digestive proteolytic enzymes the break down the food. These groups of Arachnida are described here: spiders, tailless whip scorpions, mites, pseudoscorpions, dinospiders, pygmy vinegaroons, scorpions, and daddylonglegs.

Spiders (Order Araneae)

With more than 46,000 species documented worldwide, spiders are the most species-rich order of arachnid. They are also one of the most abundant groups of animals—and can

Jumping spider
(*Phiale formosa*)

Golden orb weaver
(*Nephila clavipes*)

Freshly molted tailless whip scorpion *Phrynus parvulus*.

be found in virtually any terrestrial habitat. Arachnologists, biologists who study spiders, often say that no matter where you are, you are always within two meters of a spider. There is little cause for alarm, however, as most spiders are harmless. The handful of potentially harmful species are relatively rare and never aggressive.

A spider's body consists of two distinct parts (tagmata): the cephalothorax, which includes eyes, mouthparts, and legs, and the opisthosoma, which carries at its tip the silk-producing organs (spinnerets). All spiders can produce silk, but only some lineages build complex orbs, while others use silk to line their burrows or protect their eggs.

All spiders are predaceous, killing their victims (primarily insects) with a pair of fang-like chelicerae (a handful of species will also drink nectar and other plant material). Nearly all spiders produce venom, which also helps digest the tissue of their prey. Spider venom has evolved to be effective only on their intended prey (mostly insects) and thus is usually harmless to large animals such as humans (but see the next chapter for a few exceptions).

Tailless Whip Scorpions (Order Amblypygi)

Despite a ferocious appearance and a scary name, these animals are absolutely harmless. Not only do they not produce any venom, but their spiny "pincers" (pedipalps) are incapable of harming humans. But the story is different if you are a cricket or other small invertebrate. Whip scorpions, so named because of their extremely long, whip-like first pair of legs that act like antennae, are fast hunters that use their pedipalps to crush prey. They occur in great numbers in Costa Rica, where they are most often encountered on tree trunks at night, although only 4 species occur here.

Mites (Order Acari)

This large and extremely diverse group of arachnids, with over 55,000 described species, includes mostly very small animals, the majority of which go largely unnoticed in the soil, leaf litter, and on vegetation. A large number of mites is aquatic. The species of mite that we tend to notice are parasitic species such as ticks and chiggers (more about them in the next chapter). Some herbivorous species are important agricultural pests, while predatory species are used as biocontrol agents. In Costa Rica an interesting group of mites is associated with hummingbird-pollinated plants: *Proctolaelaps* and related genera feed on the nectar and pollen of flowers, and travel from plant to plant in the nostrils of hummingbirds. The use of an unwitting host to move from place to place is known as phoresy, and is a common mechanism of dispersal in mites and pseudoscorpions.

Velvet mite (Trombidiidae)

An unidentified pseudoscorpion.

Pseudoscorpions (Order Pseudoscorpiones)

Pseudoscorpions are tiny arachnids, rarely larger than 8 mm (sometimes smaller than 2 mm!), that resemble scorpions in general appearance, except for the lack of the long "tail." Their large grasping pedipalps are equipped with venom glands that these animals use to immobilize their prey, which consists of ants, flies, and other small insects. In Costa Rica, individuals of *Paratemnoides elongatus* form social groups, and they are thus able to hunt communally and subdue prey many times larger than their bodies. Pseudoscorpions colonize new areas by attaching themselves to the bodies of flying insects. Males of *Cordylochernes scorpioides* live permanently on the body of the harlequin beetle in order to meet females, who also use the beetle to move from one tree to another. Twenty three species of pseudoscorpion have been recorded from Costa Rica.

Dinospiders (Order Ricinulei)

Dinospiders, also known as hooded tickspiders, are a small group of arachnids that are inconspicuous but nevertheless have several fascinating characteristics. They date back over 300 million years to the Carboniferous period and have remained virtually unchanged. A morphological oddity is that dinospiders appear to have no head—the body ends abruptly in front of the first pair of legs and no mouthparts are visible; in fact, the mouthparts are hidden under a vertical plate (cucullus) that protects them as the animals burrow in the ground. It also helps them carry their prey, such as ant larvae, termites, and other small invertebrates. Dinospiders are common in humid lowland rainforests of Costa Rica, which has 8 reported species.

Pygmy Vinegaroons (Order Schizomida)

These small, pale arachnids can easily be confused with termites as they scurry out of sight under logs on the floor of humid forests of Costa Rica. Pygmy vinegaroons are blind and search for prey using their long, sensitive first pair of legs. At the end of the abdomen they carry a gland that can eject a stream of pungent acetic acid, though in quantities so small as to be barely perceptible to a human observer. Thirteen species of pygmy vinegaroons have been recorded from Costa Rica but little is known about their biology.

Dinospider
(*Cryptocoellus* sp.)

Pygmy vinegaroon
(*Schizomus* sp.)

Scorpion (*Tityus* sp.)

temperature, and substrate vibrations (which may indicate the presence of prey or predator). Another interesting trait of scorpions is their fluorescence when exposed to ultraviolet (UV) light. The function of this phenomenon is not entirely understood but it may be related to their ability to reflect the potentially harmful ultraviolet part of the sunlight spectrum.

Daddylonglegs (Order Opiliones)

These long-legged arachnids, also known as harvestmen, are sometimes accused of being strongly venomous. Yet it is entirely a myth as these harmless animals lack venom glands. Their only defense is voluntary autotomy—using dedicated muscles at the base of each leg, they are able to break off their limbs at will, leaving behind a confused predator. A detached leg often continues to twitch, further confusing the attacker. Daddylonglegs feed on earthworms and other small invertebrates, and some species specialize in pulling snails out of their shells using highly modified, pincer-like pedipalps. Like scorpions, members of the family Cosmetidae fluoresce brightly under ultraviolet light. Costa Rica has about 120 known species but undoubtedly more remain to be discovered.

Scorpions (Order Scorpiones)

The most feared and charismatic of all arachnids, scorpions are in fact rather harmless. With only a few exceptions, their sting, which is delivered with the sharp tip of their telson (the last segment of the "tail," or metasoma), has the potency of a bee sting. Scorpions are common in Costa Rica, where 18 species have been recorded, only one (*Tityus*) of potential danger to humans. These arachnids feed on insects and, on killing them with an injection of venom, masticate them thoroughly with their chelicerae (mouthparts), releasing digestive enzymes as they do so. Only tiny, predigested pieces of food pass into the stomach, while all indigestible elements are filtered out in the oral cavity. Thanks to their slow metabolic rates, scorpions can survive for long stretches of time without feeding, up to a year or more.

A unique morphological feature of scorpions is the presence of pectines, which are comb-like structures on the ventral side of the body. They function as sensors of humidity,

Daddylonglegs
(Cosmetidae)

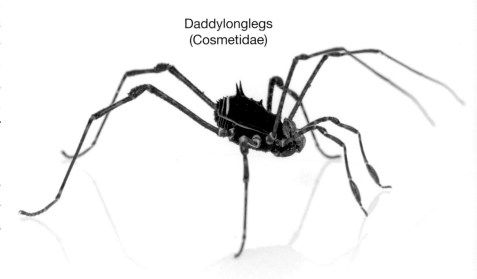

Centipedes (Class Chilopoda)

Centipedes are fast, agile predators with similar looking (metameric) segments, each carrying a single pair of legs. The number of pairs of legs is always odd, from 15 to 191, which means that a centipede can never have exactly one hundred legs. Females protect both the eggs and newly hatched young. Centipedes kill their prey, which ranges from tiny invertebrates to frogs and bats (the latter are caught in flight by giant centipedes that hang from entrances to bat colonies), using a potent venom delivered with a pair of large "fangs" (a modified pair of legs known as forcipules). Large species of the genus *Scolopendra* can deliver a painful, potentially dangerous bite, but Costa Rican species present little danger to humans. The exact number of species of centipedes in Costa Rica is unknown but is likely to be fewer than 50.

Millipedes (Class Diplopoda)

Millipedes can be easily distinguished from their relatives the centipedes by the presence of *two* pairs of legs per each segment of the body. The number of pairs ranges from 11 to 375, which is 250 legs short of the 1,000 that their common name implies. The exoskeleton is often infused with calcium salts, which increase its rigidity and strength. In addition to the protective, hard exoskeleton, most millipedes rely on chemical defenses. The majority of species produce extremely toxic compounds composed of benzoquinones, phenol, and hydrogen cyanide. Millipedes thus have few natural enemies, and those who hunt them (assassin bugs, for instance) must temporarily leave their prey behind after killing it to allow for the toxic compounds to dissipate. Many millipedes advertise their toxicity with bright coloration, but even those that have drab colors are usually left alone by birds and other predators of invertebrates. Millipedes feed on fungi, detritus, and other organic matter, and play a role in recycling dead plant material. About 100 species have been recorded from Costa Rica.

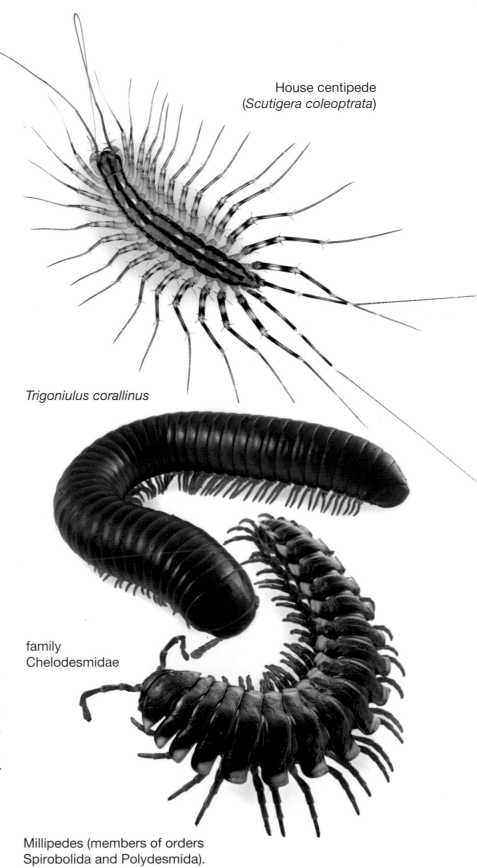

House centipede
(*Scutigera coleoptrata*)

Trigoniulus corallinus

family Chelodesmidae

Millipedes (members of orders Spirobolida and Polydesmida).

Crustaceans
(Subphylum Crustacea)

This large and extremely diverse assemblage of animals, once considered a single class, is now considered a subphylum of Arthropoda, and divided into a number of classes. Most crustaceans are aquatic, with the great majority being marine animals. A visitor to Costa Rica will certainly encounter crustaceans on the beaches of the Atlantic and Pacific coasts of the country, but it is also possible to run into an unexpected crab on the rainforest floor, far from the ocean. Several lineages of crustaceans, members of the orders Isopoda, Amphipoda, and Decapoda, have been able to conquer terrestrial habitats and evolve reproductive strategies that free them from aquatic development. Freshwater crabs (family Pseudothelphusidae) are common in the forests of Costa Rica, where they can be seen collecting fruits and hunting small invertebrates. Although they need to return to water occasionally, these crabs do not have aquatic free-living larvae. Instead, the female carries the eggs until young, fully formed, crabs hatch.

Freshwater crab
(*Potamocarcinus* sp.)

Velvet worm
(*Epiperipatus* sp.)

Velvet Worms
(Phylum Onychophora)

Although these animals are not arthropods but rather members of their own phylum, Onychophora, they are included here as they might be confused with millipedes or caterpillars. Velvet worms can be distinguished from millipedes by their soft body with no distinct segmentation, and from insect larvae by the presence of long, flexible antennae.

Velvet worms date back to the Cambrian, about half a billion years ago. They possess certain characteristics that indicate their close relatedness to arthropods, including the presence of a chitinous exoskeleton that must be periodically shed, although theirs is much thinner than that of insects. They usually reach 3–10 cm in size, but Costa Rica is home to the world's largest species, *Peripatus solorzanoi*, which can reach lengths of 22 cm.

Velvet worms have a pair of papillae, highly modified limbs situated on either side of the mouth; the papillae are connected to glands that produce extremely sticky, glue-like slime. The slime can be squirted at an incredible speed of 5 meters per second, completely entangling and immobilizing prey that is sometimes as big as the velvet worm itself. Seven species of this phylum have been recorded from Costa Rica.

2

Is It Dangerous?

It may come as a surprise but very few insect species pose any danger to humans. With the exception of a few blood-feeding insects—among them mosquitos and bed bugs—insects will always try to avoid contact with us. Like all animals, insects and other arthropods have evolved defense mechanisms that help them survive encounters with predators, but even those that can deliver a painful bite or sting will do so only if threatened or cornered. Those equipped with chewing mandibles, normally used for eating plant or animal tissue, can bite the attacker, while those that have a stinger will not hesitate to deploy it to defend themselves. Some insects have bodies covered with protective spines or bristles that can deliver a painful jab, whereas others use toxic compounds sequestered from plants to make their entire bodies unpalatable. But the one thing that insects and other arthropods will never do is to seek out and attack organisms that are not their intended prey, especially one as large and powerful as we humans.

The danger that some insects pose comes not from their bites or stings but rather from pathogens that they inadvertently transmit to us. In Costa Rica, like everywhere else, the most dangerous arthropods are mosquitos. Although insects such as Africanized honey bees or bullet ants can deliver very nasty stings, the experience is not life threatening except for someone highly allergic to bee venom or who is stung repeatedly (over six hundred times by honey bees, for example). Some Costa Rican scorpions and spiders possess potentially lethal venom, but the probability of someone being exposed to them is exceedingly low, and no fatalities from these animals have ever been recorded in this country.

The rule of thumb for dealing with potentially harmful arthropods is that if you don't bother them, they won't bother you. Needless to say, never pick up or otherwise provoke these animals, and do not approach wild beehives or wasp colonies. While enjoying an afternoon walk in the rainforest, apply insect repellent to your skin, preferably one containing DEET or Picaridin (oil of lemon eucalyptus also appears to be effective for repelling mosquitoes; citronella-based products are completely ineffective against them, and in a recent study they were shown to *attract* mosquitoes), and consider spraying your clothes with a permethrin-based tick repellent to avoid chiggers (tiny parasitic mite larvae).

But the best defense against arthropod-related dangers is knowing what can and cannot harm you, and the following section highlights the most important groups of arachnids and insects that a visitor to Costa Rica should appreciate from a safe distance.

Venomous Biters

Virtually all spiders have venom glands and are capable of delivering a painful bite if they have to defend themselves. The occurrence of spider bites is exceedingly rare—in the author's entire career as an entomologist he has never been bitten by a spider and met only one person who has (in that case, the person was bitten by a very inexpertly handled tarantula.) Nonetheless, the myth of spider bites is persistent, even among medical professionals, and spiders are frequently blamed for bacterial infections and bites delivered by other organisms. The consequences of a spider bite are usually no more serious than those of a bee sting— quickly fading pain and minor swelling. Very few species have venom that is potentially lethal to humans and even if a person is bitten by one of those species, the amount of venom delivered during the bite is too small to cause life-threatening consequences. In Costa Rica two genera of spiders may be considered dangerous, although neither is responsible for human fatalities.

Black widow
(*Latrodectus* sp.)

Black Widows

There are at least three species of black widow (genus *Latrodectus*) in Costa Rica. They all have the characteristic red hourglass pattern on the ventral side of the body and a shiny, black or brown, very round abdomen (opisthosoma). Black widows do not build concentric orbs—their webs appear tangled and are small. In Costa Rica, these spiders are found mostly in Guanacaste and Heredia provinces, but reportedly they have recently become common in San Jose, where they hide inside dark, dry structures.

Black widows feed on small insects, including ants, and their presence in houses has the beneficial effect of controlling common household pests. They are not aggressive and when threatened will fall to the ground and feign death. For this reason, bites are rare and only happen if the animal is handled or accidentally trapped against a person's skin.

The venom of black widows contains latrotoxin, a unique compound that causes a massive release of neurotransmitters (acetylocholine, norepinephrine, GABA), which causes pain, cramps, sweating, muscle spasms, and prolonged, painful contractions. In pregnant women, a black widow bite may cause preterm labor. In Costa Rica, black widow bites are extremely rare, and none has resulted in serious health effects. However, if a bite by a black widow is suspected, especially in the case of a child or a pregnant woman, the victim should seek medical attention.

Ctenid spider
(*Cupiennius getazi*)

Wandering Spider

The wandering spider (*Phoneutria boliviensis*), a large species that never builds webs, is found mostly along the Atlantic coast of Costa Rica. Although frequently declared to be the most venomous spider in the world, there are no confirmed cases of fatalities caused by this species. Other members of the genus *Phoneutria* found in South America can pose a threat, but they do not occur in Costa Rica. Wandering spiders are common on banana plantations in Limón province and thus workers there are the most frequent victims of these spiders' bites. The venom of *Phoneutria* causes dramatic changes in the victim's blood pressure, nausea, abdominal cramping, vertigo, blurred vision, and convulsions. These symptoms usually disappear within a day or two without the need for antivenin, although bite victims should seek medical attention. Wandering spiders are frequently confused with their close relatives, ctenid spiders of the genus *Cupiennius*. Ctenid spiders are common in lowland forests across Costa Rica, where they can be seen at night hunting insects and other invertebrates. These spiders are not aggressive and their bite is mild, with the pain lasting 10–30 minutes, and causing no ill effects.

Male wandering spider (*Phoneutria boliviensis*).

Stingers

Scorpions

Costa Rica is home to 18 species of scorpion, most of which are considered harmless. Scorpions deliver their venom with the sharp tip of their telson (the last segment of their "tail," or metasoma), using it both to subdue their prey and defend themselves from attackers. Envenomation by scorpions happens quite often in Costa Rica, with 200–400 cases reported each year, but no one has ever died from being stung. The venom of most common species is weak, causing relatively mild symptoms such as localized pain and elevated blood pressure. Only members of the genus *Tityus*, which includes large, mostly arboreal, scorpions, deliver a sting that is potentially life threatening. Although no fatal stings by these scorpions have ever been recorded in Costa Rica, they are responsible for human fatalities in Panama and several South American countries. Therefore, even though a scorpion sting is unlikely to cause a serious health emergency, medical attention following a sting is recommended.

The tip of the metasoma ("tail") of the scorpion *Centruroides limbatus*.

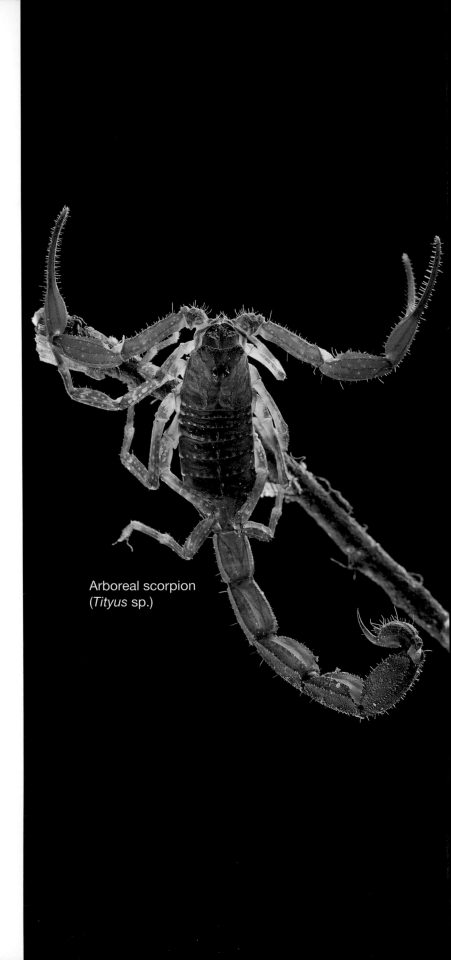

Arboreal scorpion (*Tityus* sp.)

Bullet Ant

Of all Costa Rican stinging arthropods none is more famous and feared than the bullet ant or, in Spanish, *bala* (*Paraponera clavata*). This is one of the largest ants in the world, with a body length of 30 mm, impressive for an ant. This insect is common in lowland forests of Costa Rica, absent only from the southern parts of Puntarenas Province. These carnivorous ants can be seen patrolling the forest floor and lower tree branches during the day, looking for caterpillars and other insects, which they kill with the combination of their powerful mandibles and a lethal stinger.

Their sting is considered to be one of the most painful among all stinging insects, earning them their name. According to entomologist Justin Schmidt, who created a 4-point, sting-pain index, the bala ant earns the top rating, with its sting described as "Pure, intense, brilliant pain. Like fire-walking over flaming charcoal with a 3-inch rusty nail in your heel."

The venom of bala ants contains poneratoxin, a unique paralyzing compound that directly affects the central nervous system, causing slow, long-lasting, and incredibly painful muscle contractions. Unlike the pain that follows the sting by other insects, which subsides rapidly, the aftermath of a bala ant sting is a debilitating pain that often lasts more than 24 hours. Thankfully, these ants are not aggressive and will never try to attack a person unless captured or actively threatened. The sting of the bala ant leaves no lasting negative health effects, other than a lifelong fear of large, black ants.

Africanized Bees

African honey bees (*Apis mellifera scutellata*) were imported into Brazil in 1957 in an effort to create a bee hybrid better adapted to the tropical climate of South America. They were crossbred with European bee strains, creating an "Africanized" variety. The hybrids then quickly spread across northern South America and Central America, eventually reaching the United States. Africanized bees are now common in Costa Rica, where they occur in drier parts of the country, in rocky terrain or in hollowed out trees. These insects are highly aggressive, attacking intruders at the slightest provocation. The buzz of a bee colony can often be heard from a distance and if detected should never be approached. Unlike domesticated bees, the Africanized hybrids are known to chase a victim for hundreds of meters. Although their venom is no more potent that that of a domesticated bee, the sheer number of stings that an angry swarm can deliver may be lethal to a person. If attacked by a bee swarm the best response is to run, in a straight line, as fast and as far as possible, and seek shelter in a building or a car. Protect your face and airways, and cover yourself with whatever piece of clothing is available. Do not jump into water as the bees will wait for you to emerge to take a breath. Once in a safe place, remove the stingers embedded in your skin as they will continue pumping venom. If the victim of the bee attack shows symptoms of a serious allergic reaction (large hives over the body, trouble with breathing, and bouts of fainting), this person must receive immediate medical help.

Parasa joanae

Unidentified Limacodidae.

Stinging Caterpillars

Caterpillars of many moth species, especially in the families Limacodidae, Megalopygidae, and Saturniidae, defend themselves with sharp, urticating setae (hairs and spines). Some of their setae are solid and covered with microscopic barbs that can become embedded in the skin of someone who comes into contact with them, causing irritation to the skin. Other setae are hollowed out tubes connected to a venom cell. The tips of these setae break off easily if touched, releasing the venom into the skin and causing pain similar to that of a bee sting. Usually the effect of a caterpillar sting subsides within a few minutes, but in some cases the pain, accompanied by nausea, may last for a day or two. The sting usually happens on a walk through dense vegetation, when the hapless hiker accidentally brushes against a caterpillar sitting on a plant. If the urticatings hairs are embedded in the skin, they should be removed with tweezers or by the application of adhesive tape. Applying an antihistamine or a steroidal cream (such as Benadryl® or hydrocortisone) to the site of the sting often helps relieve the symptoms. Caterpillar stings usually do not require medical attention, unless the victim shows unusually severe symptoms or the setae embed in or around the eyes.

Acharia apicalis

Skin Parasites

Chiggers

Almost any visitor to Costa Rica who enjoys hiking will sooner or later become familiar with the feeding habits of chiggers. These microscopic, six-legged larvae of mites of the family Trombiculidae are skin parasites that cause intense, very unpleasant, itching and inflammation. You usually come into contact with them while walking through grassy vegetation. Chiggers concentrate around the ankles or the belt area, anywhere that clothes are pressed tightly against the skin. Chiggers do not burrow into the skin and do not feed on blood. Rather, they insert their piercing mouthparts into the skin, and their saliva, which contains digestive enzymes, dissolves the dermal cells, allowing the mites to feed on extracellular fluids.

Chiggers in Costa Rica are not known to transmit any diseases, although in other parts of the world they are vectors of scrub typhus. To protect yourself from chiggers, use DEET-based insect repellent on your skin and spray your clothing with products containing permethrin such as Permanone® Tick Repellent. Sulfur powder is also reported to be effective against these mites.

Chigger.

Bot Fly

As if a mosquito bite were not annoying enough, its aftermath sometimes results in an unpleasant surprise—the parasitic larva of a bot fly (*Dermatobia hominis*) living in your skin. Bot flies are parasites of primates and other large mammals, and humans often find themselves hosting these insects.

While in the skin, the bot fly larva, known in Costa Rica as *torsalo,* does relatively little damage, and in most cases the experience of having a bot fly is painless. The larva not only releases antibiotics that prevent the wound from becoming infected but it also produces analgesics that reduce the pain from something feeding on the living tissue.

Although the bot fly is a benign parasite that does not transmit any diseases or cause lasting damage, most people are disgusted by the thought of carrying a parasite. The best way to protect yourself is to use a DEET-based insect repellent regularly. But if you do find yourself with a bot fly in your skin, it can be removed. There is no point trying to squeeze the larva out or pull it out by holding the larva with forceps, as its body is covered with sharp, backward-facing spines that prevent the larva from being pulled out easily. The best method is to use a suction extractor pump such as a Sawyer Extractor™, which will enlarge the opening of the wound and allow the larva to be pulled out without damaging it (if only a portion of the bot fly body remain in your body, it will rot and become infected). Other methods include covering the wound with Vaseline, adhesive tape, and even raw steak, in an attempt to cut off the oxygen and force the larva to crawl out on its own. As the doctors and nurses in Costa Rica are very familiar with this parasite, the best course of action is to contact a medical professional to help remove it.

The life cycle of a bot fly is a fascinating one. Rather than risking being swatted while laying eggs on the skin of her intended victim, the female bot fly catches a mosquito and lays eggs on its body. The mosquito is then released unharmed and it continues to seek its own victims, which often happen to be people. Once the mosquito lands on a person, the body heat of the victim prompts the tiny bot fly larva to drop onto the skin and crawl into the hole made by the mosquito's piercing mouthparts. Once inside, the larva begins to feed on the skin tissue and grows rapidly. After two months it reaches the size of a peanut and eventually falls out of the skin of its host, buries into the soil, and turns into a pupa. In a couple of months an adult fly emerges and the cycle starts anew.

The adult (left), and the first and third (right) larval stages of the bot fly on the finger of their victim.

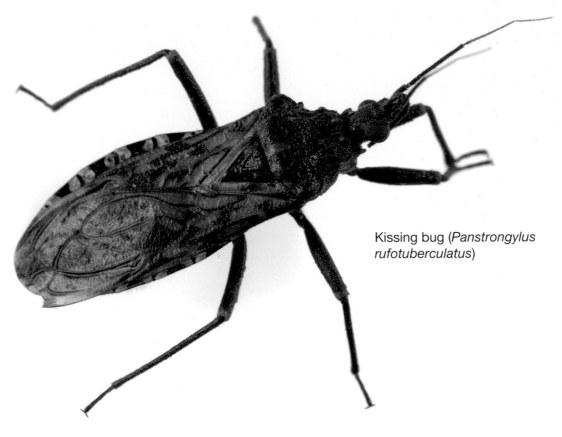

Kissing bug (*Panstrongylus rufotuberculatus*)

Disease Carriers

The most dangerous insects in Costa Rica are those that act as vectors of pathogenic microorganisms. Several groups of insect feed on human blood and in the process may infect us with viruses, bacteria, or protozoans that cause serious, sometimes life-threatening, diseases. The best way to avoid begin bitten is to routinely use insect repellent and limit your exposure by avoiding places where they are likely to be found; it is also a good idea to stay indoors at sunset and other periods of peak activity of disease-carrying insects.

Kissing Bug

Chagas disease is caused by the parasitic protozoan *Trypanosoma cruzi*. Its vectors are blood-feeding triatomine bugs (family Reduviidae) that pick up the protozoan when they drink the blood of animals or people that are already infected, and transmit it to the next victim through their feces. Triatomine bugs are found in houses with roofs made of straw and palm thatch, and thus are most commonly found in rural, poorer areas of Central America. In Costa Rica triatomine bugs are most common along the Pacific coast and in Guanacaste. People get infected with chagas disease when the bugs feed on their blood at night and subsequently defecate on their skin. The insects often feed on the face of the victim, especially around the mouth, earning them their common name.

Chagas disease can become life threatening if left untreated, and may become a life-long, chronic illness. If a bite by a kissing bug is suspected, especially if a person is experiencing symptoms such as unusual fatigue, headache, fever, or rashes, medical help should be sought immediately.

Sand Flies

Many lowland areas in Latin America, especially those adjacent to sandy river banks, are home to tiny, innocent looking sand flies of the genus *Lutzomyia*. They are vectors of a dreaded tropical disease, leishmaniasis, caused by *Leishmania* protozoans. Leishmaniasis causes deep, extensive skin lesions (cutaneous leishmaniasis) that are slow to heal and leave behind unsightly scaring. In extreme cases this may extend to mucus membranes of the nose and throat (mucucutaneous leishmaniasis). The lesions are usually painless but may become infected with bacteria, causing pain and further complications. To protect against leishmaniasis, use a DEET-based insect repellent; sleep under a net treated with permithrin; wear permithrin-treated clothing; and avoid outdoor activities between dusk and dawn, especially near rivers.

If after a trip to Costa Rica or another Latin American country you notice a small, painless wound that refuses to heal, consult a tropical medicine specialist. The treatment will depend on the species of *Leishmania* that has caused the infection.

Mosquitos

Mosquitos are the insects that cause the most fatalities and suffering across the world. They are vectors of some of the most serious diseases, including malaria, yellow fever, dengue, and many others. For many of these diseases, vaccines either do not exist or are of limited effectiveness. In Costa Rica, thankfully, some of the most serious mosquito-borne diseases such as yellow fever are absent. The risk of malaria is virtually non-existent here and prophylaxis against it is not necessary. But there are other risks from mosquito bites and the best way to limit them is to avoid being bitten. In addition to using repellent, wear neutral-colored (beige, light gray), long-sleeved clothing, and avoid outdoor activity at dusk, when mosquitos are also the most active. If possible, sleep under a mosquito net.

The most serious mosquito-borne disease in Costa Rica is dengue fever, a viral infection caused by several virus types of the family Flaviviridae. Dengue is transmitted by the Egyptian mosquito (*Aedes aegypti*) and Asian tiger mosquito (*Aedes albopictus*). These mosquitos thrive in areas with poor sanitation and standing water such as puddles, water tanks, and water collected in old tires. Symptoms of dengue include high fever, pain behind the eyes, joint and bone pain, and severe headache. As there is no antiviral treatment available, the best you can do is to treat the fever and other symptoms. The same two mosquito species are also vectors of two newly arrived diseases that have recently spread from Africa to the New World. The chikungunya virus causes symptoms similar to those of dengue fever and, similarly, there is no vaccine or antiviral treatment against it. In most cases the disease is not lethal, although it may cause serious health complications. The zika virus, another recent arrival from Africa, may cause serious neurological complications in adults and microcephaly in infants born to pregnant women infected with the virus. In most healthy adults, however, the infection is asymptomatic or exhibits itself by mild fever and muscle pains that usually disappear after a week.

Feeding Egyptian mosquito (*Aedes aegypti*), left.

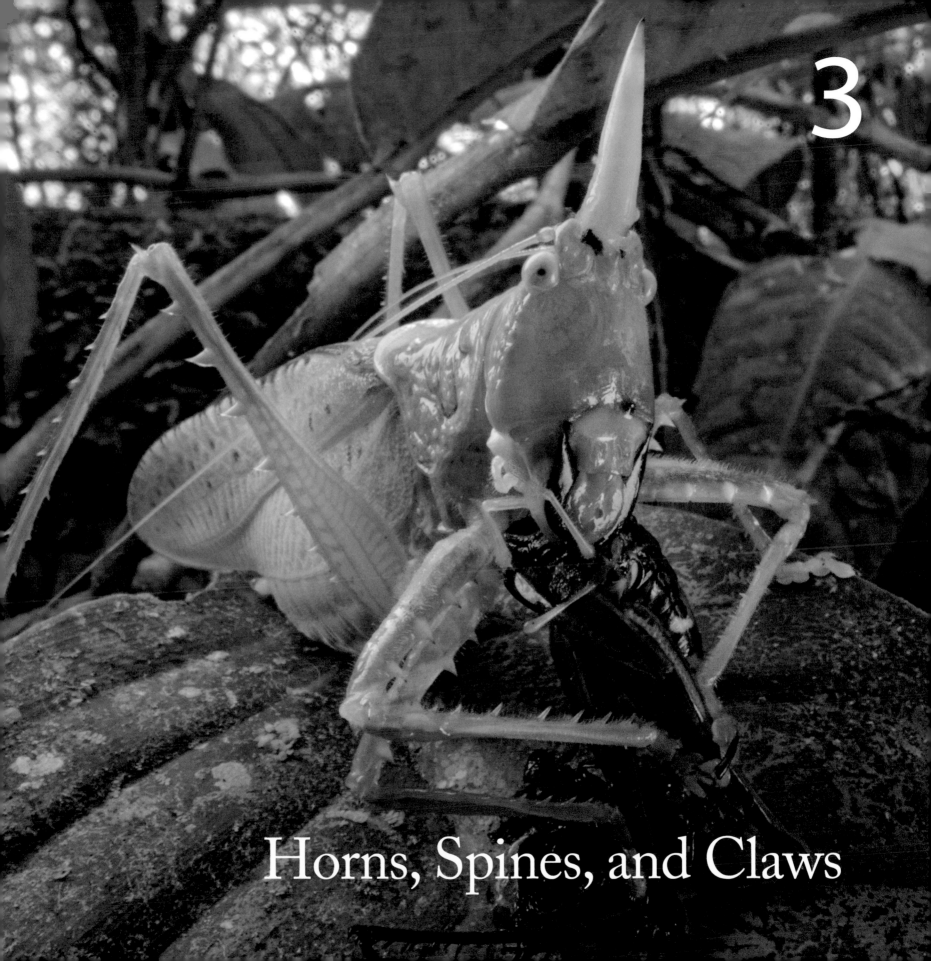

Horns, Spines, and Claws

Hercules beetle
(*Dynastes hercules*)

Among the potentially harmful insects and arachnids that occur in Costa Rica, few appear threatening. In fact, there seems to be an inverse relationship between the size and armament of an insect and the harm it can inflict on a person. A night walk in the rainforest is the best way to test this principle. If you are lucky, you will see arthropods with impressive external weaponry. While the size of their horns, spines, and claws is often an indication of their strength, these weapons are rarely used against anything other than their intended targets, which include rivals within the same species, predators, and prey. No matter how large and scary the insect might seem, rest assured that it has no intention, nor the capacity, usually, to harm a human.

Two powerful evolutionary forces govern the origin and development of external weaponry in insects and other animals. Charles Darwin was the first to point out that small differences among members of the same species have an effect on their survival and, ultimately, their reproductive success. Those individuals whose features (phenotype) lead to reproduction are able to pass these characteristics on to the next generation, while individuals with inferior features die before being able to breed. Over many generations the more successful phenotypes become dominant and refined. Darwin coined the term *natural selection* to describe this process (he was unaware of the existence of genes but intuitively understood the underlying mechanism of heredity). Darwin also noticed that in a world where a single male can mate with more than one female, some males may be left without a mating partner, creating competition among males for access to females. This, in turn, leads to the development of characteristics that give some individuals an advantage over others in gaining access to mating partners. Such an evolutionary mechanism is known as *sexual selection*.

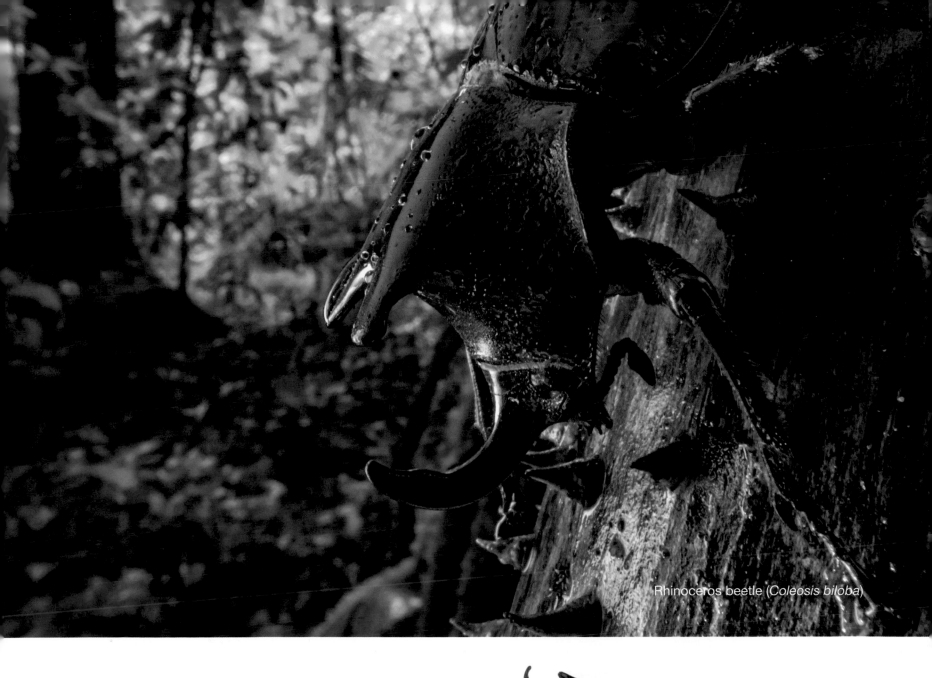

Rhinoceros beetle (*Coleosis biloba*)

Rhinoceros beetle
(*Heterogomphus mniszechi*)

female

male

Both natural selection and sexual selection can create in insects fantastic armaments, some of the best examples of which are found among Costa Rican beetles. Chief among them is the Hercules beetle (*Dynastes hercules*), the heaviest and arguably the most impressive insect in the country. In this species males fight among each other for access to females, using the long processes on the head and thorax like powerful pliers to lift, squeeze, and then toss opponents to the ground (females have no such ornamentation). Related species of rhinoceros beetle (subfamily Dynastinae) have similar weapons that have evolved under the pressure of sexual selection; species with shorter horns use them like a pitchfork to lift and twist opponents off tree trunks during fights.

Rhinoceros beetle (*Heterogomphus mniszechi*)

Hercules beetle (*Dynastes hercules*)

In dobsonflies (*Corydalus*), sexual selection has resulted in males having long, sharp mandibles. At first glance, these insects, which often come to house lights at night, can look scary. But, like so many seemingly dangerous insects, a male dobsonfly could not hurt anyone even if he tried. The gigantic mandibles are for show only, and the animal barely has enough muscle power to open and close them (biting is completely out of the question). They use these ridiculous implements in largely ritualized combat with their rivals; this is a slower, weaker version of the jostling display seen in Hercules beetles. There is also a heavy price that male dobsonflies pay for having such exaggerated mandibles. Since they cannot use them for feeding and can thus only drink water, they die within a few days.

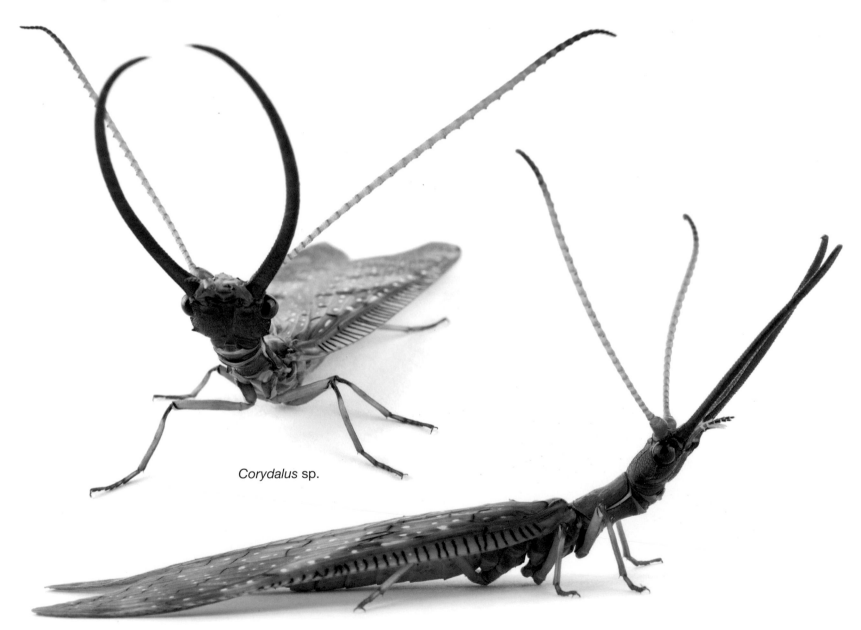

Corydalus sp.

The large pincers (cerci) at the end of the abdomen in earwigs serve a function similar to that of dobsonflies' mandibles, with male earwigs often using them to fight rivals. But while sexual selection has shortened the lifespan of male dobsonflies by making their mandibles non-functional, earwigs' cerci most likely evolved under forces of natural selection that favored individuals who can also use them for hunting and defense against predators. Female earwigs also have cerci that they use for hunting and defense, although their cerci are usually smaller than those of males. In addition, both sexes use their impressive weapons as an instrument to help them fold up, in a very intricate manner, their large hind wings. Despite being almost as long as their entire body, their folded hind wings can be hidden under the tiny, protective front wings.

Earwig (*Spongiphora* sp.)

Longhorn beetle *Derobrachus longicornis*.
Note the strongly developed mandibles.

Longhorn beetles (family Cerambycidae) are recognizable by their long, thick antennae, which are often significantly longer in males than in females. Like the armature of rhinoceros beetles and the mandibles of dobsonflies, the antennae of longhorn beetles are subject to strong sexual selection. Males with longer and more symmetrical antennae are more likely to win fights with other males to gain access to females. Their long antennae are full of chemoreceptors that allow them to detect female pheromones and potential predators. Many species in this family have strongly developed mandibles that they use to chew through the trunks of trees inside of which these insects develop. And these mandibles can cause serious damage to a predator. A spiny thorax, combined with sharp mandibles, makes longhorn beetles difficult prey for all but the strongest of predators.

Taeniotes scalatus

The Costa Rican rainforest is an inhospitable place if you are a grasshopper. Even what looks like an innocent green leaf may turn out to be a predator. Rhinoceros katydids (*Copiphora rhinoceros*) are superb stealth hunters of insects and other small organisms—they can overpower small lizards and frogs, too. Their weapons—razor sharp mandibles and muscular, spiny legs—allow them to capture and crush prey, and to defend themselves in effective fashion. Rhinoceros katydids are able to twist their legs backward over their body and grab attackers, bringing them within reach of their killer mouthparts. But there is one predator that even rhinoceros katydids are wary of; leaf-nosed bats (Phyllostomidae) are flying assassins that silently stalk insects from the air, grabbing them by the head and dispatching them with a single bite. To defend themselves from bats, rhinoceros katydids have evolved a large, needle-sharp horn on their head, which makes it impossible for a bat or any other predator to catch them.

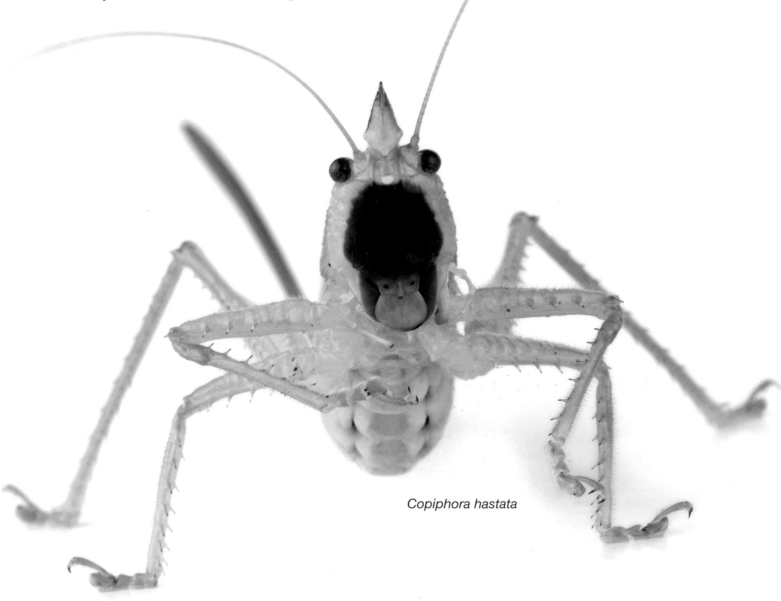

Copiphora hastata

Copiphora rhinoceros

Females of rhinoceros katydids (top) and conehead katydids (*Copiphora hastata*) carry a long, spearlike ovipositor that many people confuse for a stinger. But this harmless appendage allows females to lay eggs deep under the protective layer of dead leaves and other organic matter accumulated at the base of palm trees, their favorite place for oviposition.

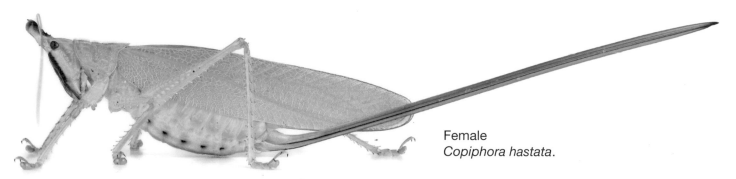

Female
Copiphora hastata.

Carrying sharp spines on your body is an effective defense against predators that swallow their prey whole, such as birds and frogs. Rhinoceros katydids are not the only animals employing this defense. Some spiders, such as the arrow micrathena (*Micrathena sagittata*), opposite, common in lowland rainforests of Costa Rica, carry large spines on their opisthosoma (abdomen) that make it virtually impossible for a predator to swallow them.

Micrathena sagittata

Fulgora laternaria

Without question, the lantern bug (*Fulgora laternaria*) is one of the most unusual insects to occur in Costa Rica. Using their long, syringe-like mouthparts, these distant relatives of cicadas and aphids feed on the sap of several species of rainforest tree. They can often be found sitting on tree trunks several meters above the ground. Lantern bugs are large insects, about 90 mm long, with a striking, blunt horn on their head.

When lantern bugs were first discovered by European naturalists, they assumed that the function of the massive horn was to produce light, though this turned out not to be the case. Today, some entomologists speculate that the main function of this hollow, balloon-like structure is to imitate the head of a lizard. Being confused with a lizard may protect lantern bugs from birds and other predators that specialize on insects. At the same time, they remain unattractive to predators that hunt true lizards, such as snakes, which use their sense of smell to detect their reptile prey. An alternative hypothesis is that the purpose of the large, brightly colored structure is to entice predators to strike there instead of at the insect's vital organs. Whatever the case, the main line of defense of a lantern bug is crypsis, simply being able to blend into the background. Its light, mottled coloration is very similar to that of the bark of trees on which it feeds. The lantern bug has yet another trick up its sleeve; if given an exploratory peck on the head by a curious bird, the insect suddenly opens up its wings to reveal bright "owl's eyes," which is bound to give the potential predator the scare of its life. The very next second the lantern bug's powerful jumping legs send it up into the air, out of harm's way.

Fulgora laternaria

Although lantern bugs are completely harmless to humans, myths about their lethality persist among some people in Central and South America. Known in Costa Rica as *la machaca*, the lantern bug is said (by the ill-informed) to be capable of delivering a bite, despite the fact that its mouth-parts are thin and flimsy. According to local lore, someone bitten by a lantern bug will die unless he or she has sex within 24 hours of the event. It is almost certain that most victims of machaca's deadly bite are young men without girlfriends.

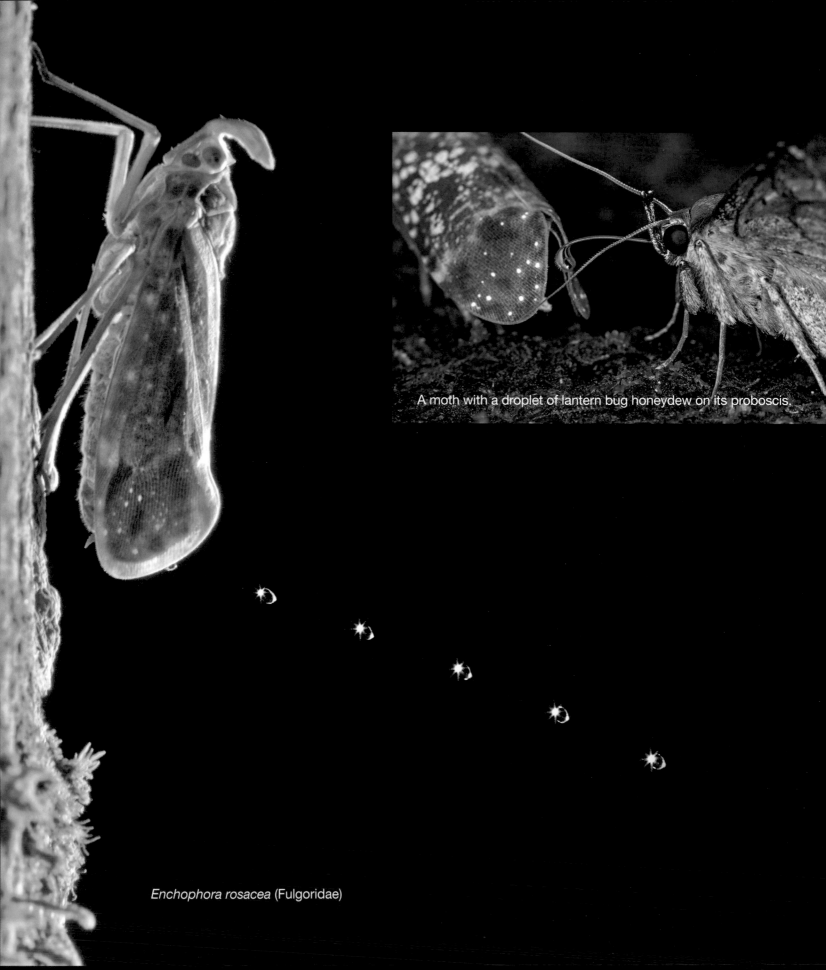

A moth with a droplet of lantern bug honeydew on its proboscis.

Enchophora rosacea (Fulgoridae)

Lantern bugs and their relatives (family Fulgoridae) have long, syringe-like mouthparts that allow them to pierce bark and tap into phloem, juices in the plant's vascular tissue. This liquid is exceptionally rich in sugars, but—of greater interest to the insects—it also contains small quantities of proteins and minerals. Through imbibing large quantities of phloem, lantern bugs end up with a surplus of water and carbohydrates, which they need to get rid off; they do so by evacuating copious amounts of a sugary liquid known as honeydew. This nutritious liquid is sought after by ants and other insects. While relatives of the lantern bug—aphids and treehoppers—produce honeydew in the form of slowly accumulating droplets, lantern bugs shoot honeydew out of their abdomen as a jet of tiny droplets that fall a great distance away from their bodies. This makes it very difficult for many insects to take advantage of this resource.

Nonetheless, several groups of animals are able to catch the flying droplets of lantern bugs' honeydew. Moths of the families Tortricidae and Noctuidae, for example, frequently visit lantern bugs, and skillfully catch honeydew with their proboscis. Another, rather surprising, animal that turns out to be an aficionado of lantern bugs' honeydew is the predatory snail *Euglandina aurantiaca* (right). The snail uses its tentacles to gently tap the lantern bugs wings, to which it responds by sending a stream of droplets directly onto the snail's body. In some instances, ants take advantage of this behavior by climbing onto the head of the snail and stealing the honeydew directly from its mouth.

Lantern bug *Phrictus quinquepartitus* (Fulgoridae) and snail.

Reaching a body length of 160 mm, the Champion's praying mantis (*Phasmomantis championi*) is Costa Rica's largest carnivorous insect. It is a silent, stealthy hunter, capable of killing some of the fastest and best defended prey, including rhinoceros katydids, themselves quite formidable predators. A praying mantis' body is the pinnacle of adaptation to a sit-and-wait hunting lifestyle. The key element is the first pair of legs, which form a powerful raptorial mechanism. The femur and tibia of each leg are armed with sharp spines that impale and immobilize the victim, while the long, muscular hips (coxae) provide extra reach. The legs are attached to the front of the pronotum, the first, strongly elongated segment of the thorax. This not only extends the range of the strike but also removes the struggling victim trapped in the raptorial legs further away from the vital organs of the mantis' soft body.

The mantis' head is capable of moving in all directions; this ability—combined with its large, extremely sensitive eyes—gives the insect an almost 360° view of the world. A mantis is one of few insects that appears to return a person's gaze, which adds to the human-like appearance of these animals. But in reality the praying mantis' compound eyes are poor at detecting static objects. Rather, their multifaceted composition is superbly adapted towards detecting motion. For this reason, a praying mantis is much better at snatching a buzzing fly out of the air than finding a caterpillar resting right in front of it on a branch.

Like many species in this group of insects, the Champion's praying mantis is highly sexually dimorphic. Females are much larger than males and their wings are short, rendering them flightless, while males are slender and fully winged.

In the complex, multidimensional universe of the Costa Rican rainforest, the male finds the female by following her pheromonal trail and, once located, he must approach the female carefully to avoid becoming her next meal. Although sexual cannibalism in praying mantids is not as common as generally believed, if a female has not fed recently she is likely to devour the male during mating. While cruel in anthropomorphic terms, the male's body will provide the material and energy to produce a clutch of eggs, ensuring that his genes pass on to the next generation.

Champion's praying mantis (*Phasmomantis championi*) devouring a rhinoceros katydid (*Copiphora rhinoceros*).

Whipscorpion (*Phrynus* sp.) devouring another member of its own species.

Of all the animals a visitor to a tropical rainforest in Costa Rica might encounter, few evoke a more visceral, negative reaction than the tailless whipscorpions. Members of the order Amblypygi, they are probably the least deserving of fear among all arachnids. They are completely harmless to us, producing no venom or toxins, and are incapable of biting, stinging, or injuring a person in any way. But whipscorpions do have a beautifully symmetrical, armored body that for some reason our psyche automatically associates with something bad or scary. This is unfortunate, because these animals are truly fascinating.

Whipscorpions belong to an ancient lineage that goes back to the Carboniferous and they have remained relatively unchanged in their morphology. Like most arachnids, these animals are predators. These sit-and-wait hunters detect approaching prey with their first pair of antenniform legs, which come equipped with an array of motion- and scent-sensitive trichobothria, bristles, and slit organs. They eat other insects mostly, but also occasionally prey on smaller member of their own species. Once the prey is detected within striking distance, the whipscorpion pounces on it and crushes it with its spiny, muscular pedipalps (modified mouthparts). The prey is then masticated with smaller chelicerae (mouthparts) and digested externally, before the liquified meal is directed into the mouth via grooves at the very base of the pedipalps.

Whipscorpions are passionate animals, in both love and war. Their courtship ritual is surprisingly elaborate and gentle, and may involve hours of slow dancing, tender caressing, and hand-holding, before the male places an external spermatophore on the ground and slowly guides the female to pick it up. Later, the female produces a clutch of eggs that she carries on her body until the babies hatch. After they emerge, the young whipscorpions, known at this stage as praenymphs, are still quite helpless, and the mother continues to carry them on her body until they are ready to fend for themselves. Young whipscorpions occasionally aggregate into small groups, but adults, males in particular, are fairly solitary, territorial creatures. They do not tolerate the presence of another individual of the same sex in their vicinity, resulting in intense albeit largely ritualized battles.

Female *Phrynus parvulus* with newly hatched young on her back.

Male *Phrynus parvulus*.

Big, hairy, and agile, tarantulas (family Theraphosidae) epitomize danger to many people. Nothing could be further from the truth. These giant arachnids, common in Costa Rica, are less dangerous than a bee and you are definitely less likely to actually come into contact with a tarantula. There is no denying that tarantulas have large fangs connected to venom glands and, if provoked, can bite. But that happens rarely, as these animals prefer to rely on a line of defense that is not only less traumatic to the attacker but also less costly to the spider.

When threatened, a tarantula is likely to do three things. First, the spider will start rubbing its hind legs against the hairy abdomen. The body of a tarantula is covered with long, urticating hairs that can cause unpleasant irritation if they get into the eyes or any mucus membrane of the attacker. Second, the arachnid will rear its front legs and open its enormous fangs, which, in some species, are capable of puncturing a mouse's skull. This is often enough to make even the largest attacker back off. The bite itself is very costly to the spider since its venom contains proteins that

are difficult to synthesize, and the animal would rather use them for hunting. Lastly, many tarantulas produce a loud hissing sound. For the longest time the source of the sound was a mystery, but now we know that it is produced by setal entanglement: some of the hairs (setae) on the legs are covered with microscopic hooks that scrape against other, producing the loud warning hiss.

Costa Rican tarantulas can be found in almost every habitat. Some are arboreal whereas others prefer to occupy deep underground burrows. They generally feed on a variety of insects and other invertebrates, although some of the largest species are capable of hunting small rodents.

Brachypelma sp.

4

Masters of Deception

On entering a Costa Rican rainforest, your initial thoughts, surprisingly enough, might be that it is virtually devoid of animal life. Sure, you hear birds and cicadas high in the canopy, and you might get a glimpse of an anole in the leaf litter, but where are all the insects? Other than butterflies fluttering around flowers and ants scurrying at your feet, the forest looks empty. This is an illusion, of course, an illusion perfected by millions of years of evolution. Thousands of species of insect are hiding in plain sight, displaying some of the most amazing examples of camouflage on the planet.

In addition to finding food and mates, all animals are preoccupied with one critical goal—to avoid being eaten. Some animals develop weapons and other defenses that simply make them too dangerous to tangle with. Large mammals and reptiles have few natural enemies simply because other animals rarely match their size and strength. But when you are as small

as a beetle, this option is not available. Consequently, insects and other arthropods have only two choices. One is to make themselves unpalatable, and thus of no interest to predators. The other is to hide. Some hide by squeezing under bark, rocks, and other tight spaces that make their detection difficult. Others prefer to hide through camouflage: "Move along, folks. Nothing to see here."

In the dense, chaotic environment of the rainforest, remaining motionless will often allow an animal to go undetected. Indeed, I never cease to be amazed by the fact that even a deer or other large animal will often go unnoticed simply by standing still. But a patient, deliberate predator, one that carefully scans the environment for recognizable shapes and colors, will ultimately detect the presence of its intended prey. Birds, monkeys, and many lizards are equipped with excellent vision and can recognize the familiar shape of

Sylvan leaf katydid
(*Mimetica viridis*)

insect prey, even if it is not moving. After all, a small symmetrical object with legs and wings can only be one thing. For this reason, insects that cannot hide under rocks or logs—often because their wings are too large or delicate—have had to evolve adaptations that allow them to become invisible to predators by assuming the appearance of a lifeless, inanimate background. This ability to make oneself invisible by resembling a leaf, a stick, or a splash of bird droppings is known as crypsis.

Crypsis employs several different strategies. In its simplest and most common form, the insect assumes the color of the background on which it rests. A green caterpillar on a leaf or a brown moth on the bark of a tree can attempt to blend in with the background by virtue of having a similar hue. But if identified as a separate object by a predator, then it is only a matter of seconds before the predator recognizes them as something edible. Other insects take the deception a step further by also assuming the shape of common, inanimate objects. Stick insects and leaf katydids are not only of the same color as the plants on which they feed, but their entire body bears an uncanny resemblance to a fragment of the vegetation itself. Insects whose exposed body may cast a shadow, thus revealing their position, employ the self-shadow concealment technique. Their body is often fringed with hairs, lobes, or other flat protrusions that press against the substrate, thus eliminating shadow. Lastly, many insects try to break up the easily recognizable shapes of their bodies by adopting disruptive coloration. Such coloration includes markings that create the appearance of false edges and boundaries, and hinders the detection or recognition of an insect's true outline and shape.

Slug caterpillar (*Natada fusca*)

Sylvan leaf katydid (*Mimetica crenulata*) from the lowland Atlantic forest of Costa Rica.

If you have a chance to go out into the Costa Rican forest at night, you might be rewarded with the discovery of one of the most amazing animals to occur in this corner of the world, one of the several species of sylvan leaf katydid. Don't bother looking for them during the day, however, as you will never find them—even though they will be sitting all around you! The only way to see them is to go out at night and look for the movement of their long antennae and legs as they slowly munch on leaves.

Sylvan leaf katydids, which belong to a group known as the Pterochrozini, are probably the most accomplished mimics

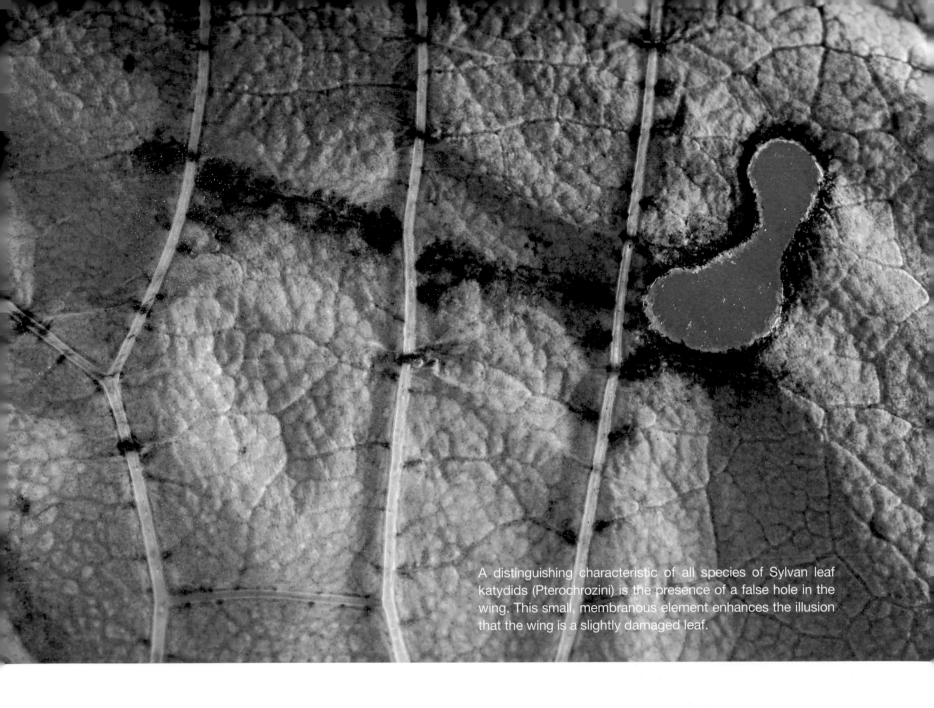

A distinguishing characteristic of all species of Sylvan leaf katydids (Pterochrozini) is the presence of a false hole in the wing. This small, membranous element enhances the illusion that the wing is a slightly damaged leaf.

of plants among all insects. Their mimicry is by no means restricted to resembling plain, green leaves—that's amateur stuff. No, their bodies are perfect replicas of leaves that have been chewed up, torn, rotted, dried up, partially decayed; sometimes, the surface of their bodies is covered by fake fungi or lichens. They even have fake holes in their wings (fake, because the holes are in fact thin, translucent parts of the wing membrane). Their mimicry is so exquisite that a famous evolutionary biologist once clipped off a part of a katydid's body to prove to me that it had real lichens growing on its abdomen. Alas, it didn't, and hemolymph was spilled.

Sylvan leaf katydid
(*Mimetica viridis*)

Nearly every individual of *Mimetica* looks different. Within the same population some insects mimic green leaves, while others perfectly resemble damaged or dead leaves.

One striking aspect of the biology of sylvan leaf katydids is that no two individuals of even the same species are alike, and within a single population you can find individuals whose appearance is so dramatically different that you would feel justified in placing them in different species. Not surprisingly, that is exactly what entomologists have done: nearly a quarter of all species of sylvan leaf katydids were "discovered" and described more than once, under different names. Such diversity of forms within a single species is known as polymorphism. There are plenty of plant-mimicking insects, but none show a similar diversity of forms within a single species. Why does polymorphism exist?

The answer has to do with the predators—primarily monkeys—that target these insects. Monkeys actively search for katydids, unfurling leaves and systematically combing through the vegetation. For some species of monkeys, katydids can constitute over 80% of their diet. Primates are very smart animals, and even the best mimicry would not fool monkeys if they were able to conceptualize the specific shape and coloration of, say, a species of leaf mimic. But when every individual in a katydid population looks slightly different, then the task of finding them is much more difficult. In Costa Rica, katydids of the appropriately named genus *Mimetica* show such diversity of forms that to this day entomologists are not sure how many different species of these insects exist. Within the same population, individuals that are green, brown, yellow, or partially brown and partially yellow—some smooth, others spotted –can share the same tree, with no difficulty in spotting a mate. All to fool predators.

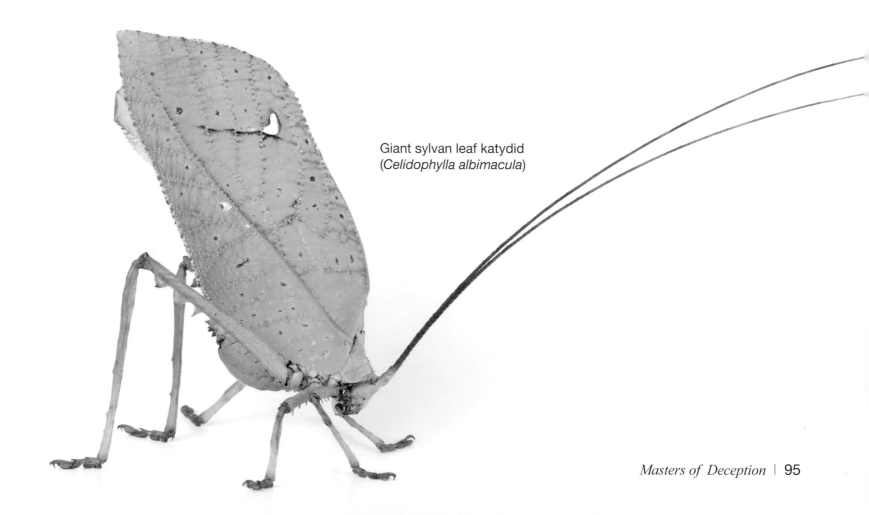

Giant sylvan leaf katydid
(*Celidophylla albimacula*)

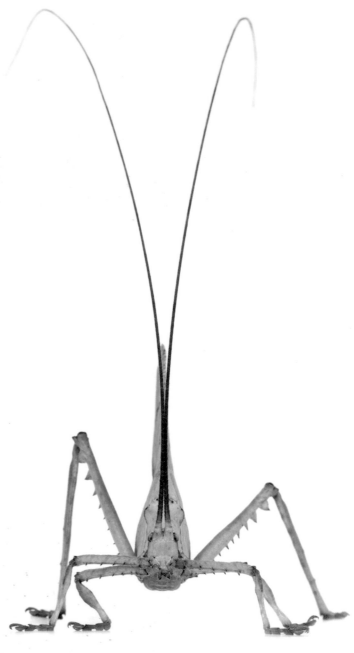

The largest Costa Rican sylvan leaf katydid is *Celidophylla albimacula* (above and left). About the size of a child's hand, this insect's perfect resemblance to a large, fresh leaf makes it disappear even as it sits fully exposed on a branch.

Spotted katydid (above and below).

The spotted katydid (*Anapolisia maculosa*) has mastered the art of disappearing right in front of your eyes. Its coloration matches the surface of a typical large rainforest leaf covered with epiphytic lichens and moss. The katydid's camouflage is exquisite; its wings not only resemble leaves with lichens, but the individual spots correspond to the common lichen genera *Gyalectidium* and *Calopadia*. At the same time, the partially translucent wings allow the light to shine through, further obliterating the familiar shape of this insect.

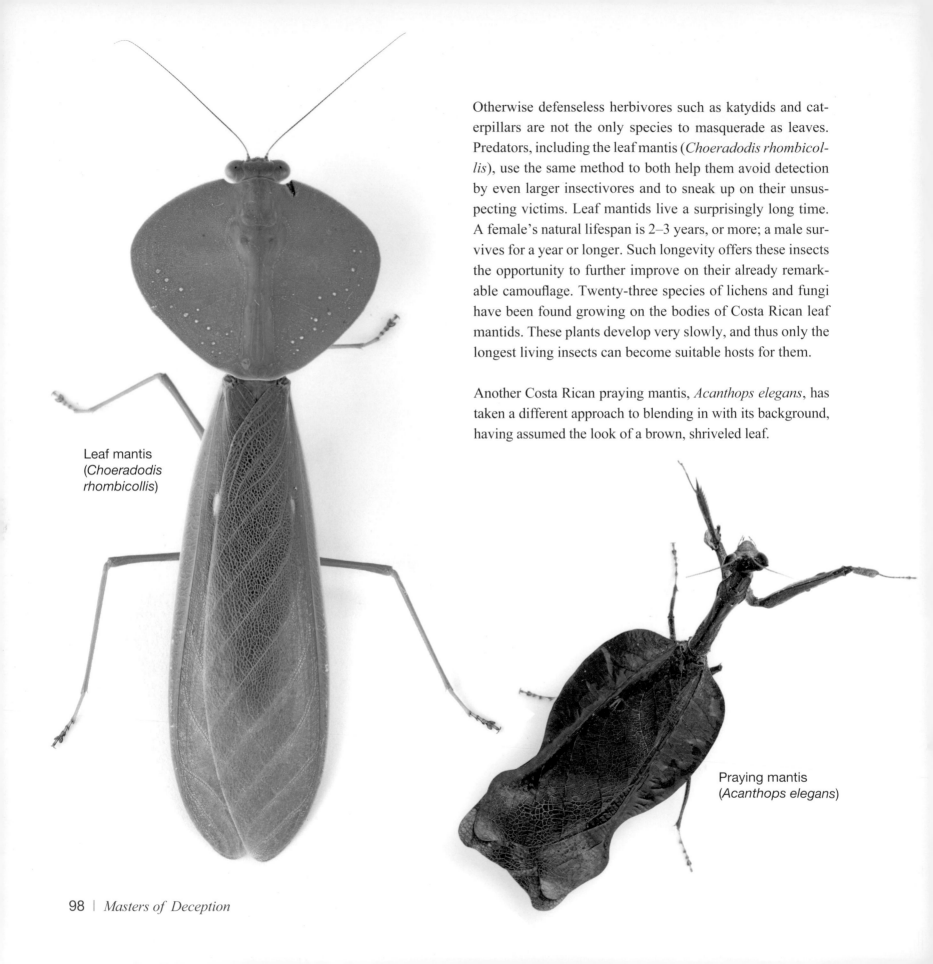

Otherwise defenseless herbivores such as katydids and caterpillars are not the only species to masquerade as leaves. Predators, including the leaf mantis (*Choeradodis rhombicollis*), use the same method to both help them avoid detection by even larger insectivores and to sneak up on their unsuspecting victims. Leaf mantids live a surprisingly long time. A female's natural lifespan is 2–3 years, or more; a male survives for a year or longer. Such longevity offers these insects the opportunity to further improve on their already remarkable camouflage. Twenty-three species of lichens and fungi have been found growing on the bodies of Costa Rican leaf mantids. These plants develop very slowly, and thus only the longest living insects can become suitable hosts for them.

Another Costa Rican praying mantis, *Acanthops elegans*, has taken a different approach to blending in with its background, having assumed the look of a brown, shriveled leaf.

Leaf mantis
(*Choeradodis rhombicollis*)

Praying mantis
(*Acanthops elegans*)

Leaf mantis (all three photos).

The tree mantis (*Liturgusa* sp.) has excellent vision.

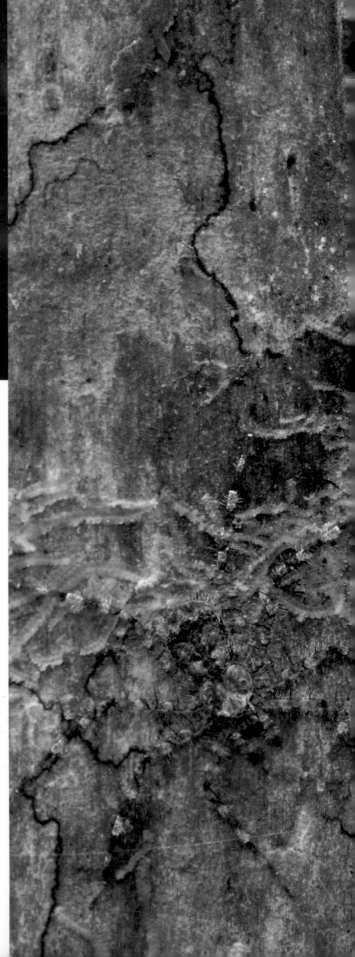

For small, flat-bodied predators, there is no better place to conceal themselves—from predators and prey—than on vertical tree trunks covered with lichens and moss. Within this environment hunt several groups of arthropods, among them insects that frequently assume the appearance of just another flake of lichen-covered bark. The tree mantis (*Liturgusa*) is a blazing fast hunter of smaller insects. Its excellent vision allows it to spot its prey from a distance and chase it down before the victim realizes that it is being hunted. But another, even better concealed, predator is a potentially mortal enemy of the mantis. The crab spider (*Selenops*) has the advantage of possessing a flatter body and legs that are fringed with long hair, a combination that allows it to completely avoid casting a shadow. This, in turn, lowers the chance of being detected by a bird or other predator, and lets it sneak up more easily on victims that have good eyesight, such as the tree mantis.

Crab spider (*Solenops* sp.) and tree mantis (*Liturgusa* sp.).

Hameodiasma tessellata (above and left).

Species that employ camouflage as their main line of defense must display a great degree of polymorphism. This prevents predators from developing a search image that will allow them to decipher the insect's camouflage—but it also serves another purpose. The tessellated sylvan katydid (*Hameodiasma tessellata*) is a species found in Costa Rica in habitats that range from hot, humid lowland rainforests to cold, misty cloud forests at elevations of 2,400 m. At lower altitudes, this katydid prefers to hide on vibrantly green vegetation and moss-covered tree trunks, and its coloration matches such background. But in the colder environment of Costa Rica's Talamanca Mountains, the same species becomes a replica of tree bark covered with splotches of white lichens, typical of habitat at higher elevations.

The closely related Champion's katydid (*Championica montana*) takes advantage of the ubiquity of moss in the rainforest, having evolved coloration and ornamentation that allow it to blend seamlessly into thick carpets of moss. Unlike leaf- and bark-mimicking katydids that often achieve complete crypsis through coloration alone, moss-mimicking forms have adapted nearly every element of their bodies to create the illusion of fragmented, small-leaved vegetation. Thick spines on their legs and thorax not only help mask the outline of the insect's body but also provide excellent defensive weaponry. If detected by a predator, this species viciously kicks its legs, fans its strikingly colored wings, and raises its abdomen to reveal bright coloration that warns of severe consequences to anyone trying to swallow the seemingly defenseless animal.

Championica montana

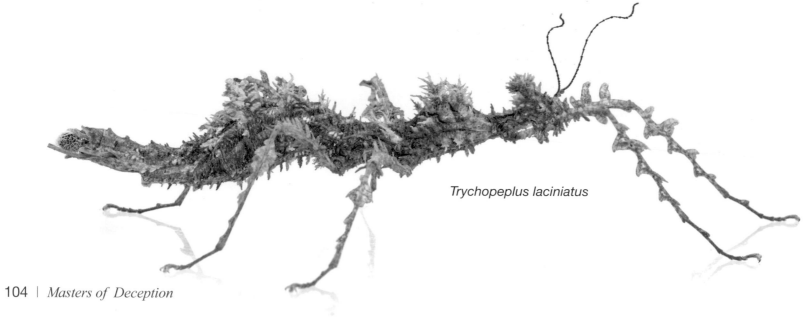

Trychopeplus laciniatus

Rhynchacris bigibbus

Prisopus sp.

Walking sticks (order Phasmida) are masters of pretending to be just another twig. Everything about their bodies is designed to make predators think that they are looking at a plant. Species that live at higher elevations, such as the otherworldly *Trychopeplus laciniatus*, have not only evolved an elongated body shape but also intricate ornamentation that resembles a branch thickly coated with moss. Others tightly hug branches to become one with them and escape detection. But in addition to visual defenses, many walking sticks employ an arsenal of chemical weapons. Little is known about the chemistry of most tropical species, although many—including members of the Costa Rican genus *Prisopus*—produce a strong odor if disturbed and are avoided by predators. A unique toxic compound, anisomorphal, was discovered in some highly noxious walking sticks. Unlike grasshoppers and other insects that feed on toxic plants and sequester their secondary compounds, walking sticks, which feed on a variety of innocuous plant species, synthesize their own chemical protection.

Walking sticks are not the only insects that are good at resembling pieces of wood. The stick praying mantis (*Angela*) rivals the thinnest and most twig-like members of the order Phasmida in its resemblance to a dead stick.

Another approach is to look like a small fleck of bark or decaying wood, and certain moth species excel at this. Members of the genus *Pentobesa* (family Notodontidae) only begin to resemble an insect if placed on a smooth, white surface, against which its legs—and identity—become obvious. But when resting on vegetation or on the forest floor during the day, it vanishes right in front of your eyes. Some longhorn beetles have adopted a similar approach by evolving coloration and texture that bears an uncanny resemblance to a piece of dead bark.

Moth (*Pentobesa* sp.)

Stick praying mantis (*Angela* sp.)

Prominent moth (*Pentobesa* sp.)

Long-horned beetle (*Lagocheirus* sp.)

Cladonota sp.

Guayaquila sp.

Some species of the treehopper family Membracidae have embraced the strategy of looking like the defensive parts of plants—thorns, spikes, and hooks. Many of these insects lead communal lives, together creating the appearance of a string of thorns on a branch. Treehoppers are fairly sedentary and rarely move while feeding, which enhances the illusion that they are indeed parts of the plant.

In addition to serving as effective camouflage, the large projections on the treehoppers' back (a highly modified pronotum, the first part of the thorax) also serve to defend themselves against their mortal enemies, wasps. Birds and other large predators are also unlikely to try their luck swallowing something that is as unpalatable as a fishhook.

Umbonia crassicornis

Weevil (*Conotrachelus* sp.)

Unidentified spider (Araneidae)

Treehoppers (*Bolbonota xalapensis*)

Although it may seem rather undignified to us, one of the best strategies among arthropods to avoid detection by predators is to look like something very few animals are interested in—bird droppings. Many species of spiders, beetles, and caterpillars sport a combination of white and brown splotches that quite convincingly resemble bird poop. This allows them to rest fully exposed and unperturbed on the surface of leaves, which are often adorned with the genuine article. Smaller insect species such as treehoppers of the genus *Bolbonota*, which are only 2–3 mm long, very convincingly mimic the appearance of caterpillar droppings (frass).

A common sight in tropical forests is bodies of insects covered by a strange growth. No longer autonomous animals but merely a substrate to nourish a foreign life form, these insects have fallen victim to a sneaky pathogen—an entomophagous, or insect-eating, fungus. Fungi like these grow inside insects' bodies and dramatically alter their behavior. In many cases the insect is driven to climb tall plants, and shortly before dying firmly attaches itself to the plant with its legs and mandibles. After the host's death, the fungus continues to grow within it, eventually producing external fruiting bodies that will disperse new spores. In an effort to appear as unappealing to predators as possible, some insects have evolved camouflage that mimics the appearance of such diseased, dead bodies. Some do it by coloration alone, others produce cuticular outgrowths that resemble the tendrils of a deadly fungus sprouting from their bodies.

An ant (left) and a moth (below)
killed by entomophagous fungi.

A butterfly chrysalis (left) and a moth (below)
mimicking the appearance of fungal infestation.

Weevil (*Heilipus* sp.)

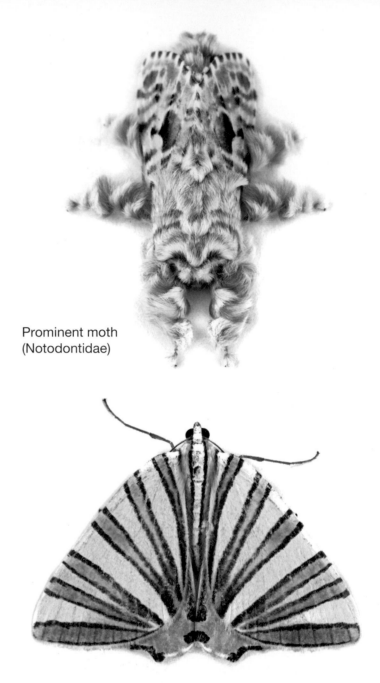

Prominent moth
(Notodontidae)

Moth (*Pityeja histrionaria*)

Sometimes the best camouflage is to simply look like something else, anything other than the intended target of the principal predator. For this reason, many insects employ disruptive coloration that creates the appearance of false edges and makes the outline of the true shape of the animal difficult to discern. This technique is common among moths, which often have wings adorned with lines and markings that distract the viewer from the real, fairly uniform, shape of these insects. Others, including many beetles and some katydids, rely on distractive patterns that direct the attention or gaze of the predator away from traits (the outline of the body, for example) that would otherwise give away the animal.

Moth (*Epia muscosa*)

5
Warning Signs

Despite displaying incredible diversity, human societies across the globe have at least one thing in common, a love of colorful objects. Papuan warriors adorn themselves with the vivid plumages of birds of paradise, Masai cattle herders use bright red cloaks and multicolored necklaces, Costa Rica's richly kaleidoscopic oxcarts are famous worldwide, and pink seems to be a favorite color on beaches worldwide. In the animal kingdom, countless bird species wear vibrant colors, while bowerbirds and some other species actively seek out and collect colorful objects to attract their mates. The brighter the male bird or his treasure trove, the greater the chances that he and not his duller rivals will be selected by a female. In other words, humans and birds seem to equate rich, polychromatic designs with beauty and other attractive qualities. But things are a little different in the insect world.

When we compare the appearance of closely related insect species that are active at night with those that go about their business during the day, the latter species will almost invariably show more distinctive color patterns. This makes a lot of sense, as the interactions between insects—be it with predators or mating partners—are often based on visual signals. While it is true that some diurnal insects use coloration to

Velvet ant (Mutilidae)

blend in with their surroundings, many others use showy coloration to stand out—and send a loud message. Occasionally this message is "come to me," as is the case among male dragonflies that flash the bright colors of their abdomens to attract females. But more often than not, the bright colors spell out "stay away or I may hurt you." A colorful insect hopes to be perceived as a dangerous insect.

At the end of the nineteenth century, evolutionary biologist Edward B. Poulton introduced the term *aposematic coloration* to describe the colorful, gaudy patterns used by insects and other animals to alert potential predators of their ability to inflict harm if attacked. Aposematic, or warning, colors are never subtle and usually rely on a simple color palette of yellow, black, red/orange, and, less often, white and blue. Why these particular colors? It seems that for many groups of diurnal predators that rely on their sense of vision—among them primates, birds, and, to a lesser extent, reptiles and amphibians—these colors are the easiest to learn, the most difficult to forget, and the most different from the typical colors of edible prey. Interestingly, human societies have adopted similar color schemes to develop a system of signals that warn of potential dangers. The yellow and black tape that says "Caution," the orange and black "Biohazard" warning notice, the red and white "Stop" sign, and the red and blue police flashing lights all stem from our innate predisposition to pay attention to such colors.

But what is the threat that insects with aposematic coloring are announcing? Generally, for the message to be effective, it must be backed up by a real, potentially harmful, defense. In virtually all cases of aposematism in insects, the weapon they use is chemical, either in the form of venom (a substance to be injected into the attacker's body) or a poison (a substance that needs to be ingested by the attacker). In some cases, the chemicals contain unpleasant smells or bitter flavors, with little or no toxicity, and merely repel the attacker. Rarely is aposematic

Tiger moth (*Dysschema leda*)

coloration used by insects whose only defense is the strength of their mandibles or the powerful kick of their hind legs.

The most familiar examples of aposematic colors are found among members of the order Hymenoptera, insects famous for carrying sharp, painful stingers that they do not hesitate to use. Wasps and bees are immediately recognizable by the presence of pale (usually yellow) and dark (brown or black) bands on their abdomen and thorax. Less frequently, stinging wasps employ other colors such as black and orange or metallic blue. Their warning message is extremely effective. A single encounter with a wasp is usually enough for any person or bird to learn a lesson that will never be forgotten— yellow and dark stripes signify pain.

Wasp (*Mischocyttarus* sp.)

Given that the hymenopteran stinger is actually a modified egg-laying organ (ovipositor), only female wasps and bees can sting. But the ability of these insects to cause pain is so strong that even entomologists, myself included, are reluctant to handle male wasps. The stinger is connected to a venom sack at the tip of the abdomen. In bees, whose stinger is barbed—which causes the sack to be ripped out of the abdomen when the stinger is used—the musculature of the sack continues pumping venom even when the sack is fully detached from the insect's body.

The venom of wasps and bees contains multiple antigens that cause severe reaction in humans and can be deadly to smaller predators. The first reaction is sharp pain that causes the predator to immediately drop the insect. The pain is caused by acetylcholine, a neurotransmitter that activates nociceptors, pain receptors in the skin. This is followed by a strong inflammatory response and widening of blood vessels (vasodilation), which increases their permeability and helps spread venom through the tissue. In most cases, the painful effect of the sting lasts only a few

A single female nest of the paper wasp (*Polistes* sp.)

Multi-female nest
of paper wasps
(*Polistes* sp.)

minutes, but the venom of certain wasps may cause sharp pain lasting several hours.

Insects that carry stingers or venomous spines (such as slug caterpillars of the family Limacodidae) are able to announce their unpalatability to a predator without being wholly ingested by the predator. The predator's brief contact with the insect's body, one that does not result in any damage to the body, is enough to imprint it with the knowledge that it is best to avoid such insects.

Wasp Mimics

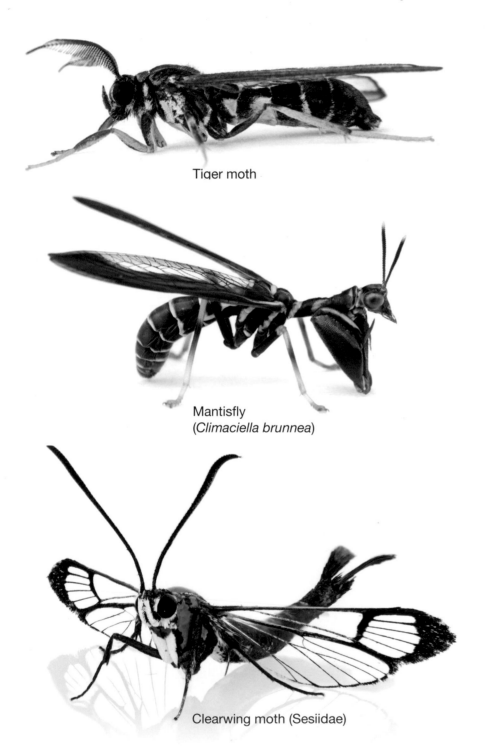

Tiger moth

Mantisfly
(*Climaciella brunnea*)

Clearwing moth (Sesiidae)

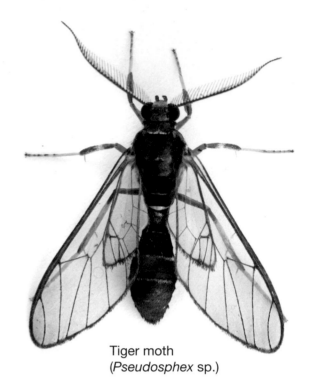

Tiger moth
(*Pseudosphex* sp.)

Tiger moth
(*Dasysphinx volatilis*)

The desire of predators to avoid the venom of wasps and bees has been widely exploited by other insects. Some species—belonging to many unrelated groups—have evolved to resemble Hymenoptera, though they lack the stinger and often any other defenses. This phenomenon of harmless species pretending to be harmful ones is known as Batesian mimicry, in honor of British naturalist Henry Bates, who was the first to document the existence of harmless, very convincing, mimics of chemically protected species.

Batesian mimicry of wasps is particularly common among moths, which usually evolve clear wings that are devoid of the scales typical in this group of insects. Some even adopt a color pattern that mimics the presence of the typical wasp "waist" that connects the abdomen to the thorax. Wasp mimics can be found in other groups of insects, including such unlikely mimics as planthoppers, lacewings, and even katydids and crickets. In addition to having the appearance of a wasp, such species adopt the behavior of wasps, with fast and jerky movements often accompanied by rapid twitching of their antennae, behavior typical of many insect-hunting wasps.

Tiger moth
(*Isanthrene* sp.)

Planthopper
(*Oncometopia* sp.)

Jabbing a predator with a painful stinger, while very effective, is not the only way insects defend themselves. In fact, the most common defenses against predators are not venoms that are injected into the body of an attacker but poisons that are delivered if a predator attempts to devour the insect. The poisons are not always designed to kill the attacker but rather to make it quickly reject its prey. This not only makes the predator immediately associate an unpleasant interaction with the insect but also minimizes the damage to the insect's body.

For this reason, many insects produce chemicals that act as emetics, causing the predator to regurgitate or drop prey before a larger, potentially lethal, dose of the toxin is absorbed. Such toxins tend to be very bitter, which further lowers the chance that the attacker will continue eating the prey. In monarch butterflies (*Danaus plexippus*), their caterpillars feed on milkweed (*Asclepias*), plants rich in cardenolide aglycones (a type of steroid), which causes birds to instantly regurgitate the butterflies and to develop a lifelong aversion to the

Monarch butterfly (*Danaus plexippus*)

Firefly (Lampyridae)

orange and black coloration displayed by this species. The bodies of fireflies (beetles of the family Lampyridae) contain lucibufagins (another steroid), which has an effect similar to that of the steroid contained in monarch butterflies.

Caterpillars of the sphinx moth *Pseudosphinx tetrio* feed on plants of the family Apocynceae, which are rich in cardiac glycosides. They are able to sequester these toxins and use them for their own defense. These caterpillars advertise the unpalatability of their bodies with bold, contrasting coloration. Some biologists speculate that their coloration mimics highly venomous Costa Rican coral snakes.

Sphinx moth caterpillar
(*Pseudosphinx tetrio*)

*Cosmosoma
teuthras*

*Cosmosoma
zurcheri*

*Gymnelia
nobilis*

Cosmosoma sp.

Tiger moths of the family Erebidae include some of the most colorful and striking insects in Costa Rica. Unlike their nocturnal cousins, tiger moths are usually diurnal and surprisingly sluggish, which makes them conspicuous and easy to catch. When caught by a predator, they often display a behavior known as cataleptic seizure, bending their immobile body to display additional bright markings normally hidden under the tufts of hair on their abdomen. Such colors are effective in scaring off predators by indicating the presence of the highly toxic compounds in their bodies. Tiger moth caterpillars frequently feed on plants rich in toxic secondary compounds, including pyrrolizidine alkaloids and cardiac glycosides; the adults of some species inherit the chemicals sequestered from these plants. In other species, the adults also collect toxic chemical compounds from plants. This behavior is particularly common in males, who use these substances not only to gain protection from predators but also as precursors to synthesize pheromones used to attract females.

Some tiger moths such as the beautifully colored *Hippocrita drucei* react to being captured by producing copious amounts of bitter-tasting, highly toxic foam.

Net-winged beetle (cf. *Calochromus* sp.)

Tiger moth
(*Correbidia* sp.)

Leaf beetle
(*Monocesta* sp.)

Net-winged beetles (family Lycidae), among the most toxic insects in Costa Rica, are avoided by predators ranging from birds to spiders to predaceous insects. The beetles are usually brightly colored, frequently displaying aposematic orange, red, and black patterns. When disturbed, they exhibit a behavior know as reflexive bleeding, producing droplets of hemolymph from pores on their thorax. Their hemolymph contains lycidic acid—a unique toxin found only in this group of insects—which is universally disliked by insect-feeding predators. In addition to lycidic acid, net-winged beetles also contain highly odorous pyrazines that make their bodies easily recognizable as distasteful even by predators with poor vision.

The appearance of net-winged beetles is mimicked by a number of insect species, including leaf beetles (Chrysomelidae) and tiger moths of the genus *Correbidia*. Both leaf beetles and tiger moths are frequently highly toxic themselves, leading to the question as to why they mimic other insects, rather than evolving their own, unique aposematic coloration?

Similar appearance of unrelated, chemically or otherwise protected species is known as Müllerian mimicry. German zoologist Fritz Müller noticed that unpalatable butterflies sometimes converge on a very similar appearance, often to the point that even a seasoned naturalist cannot easily distinguish them. The reason for this behavior has to do with statistics; the lower the number of distinct color patterns that indicate danger, the easier it is for predators to learn to identify them.

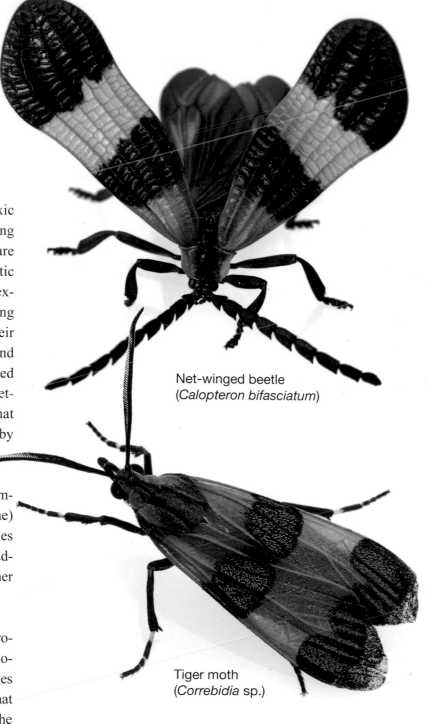

Net-winged beetle
(*Calopteron bifasciatum*)

Tiger moth
(*Correbidia* sp.)

Every four to eight years, beginning in July or August, a massive migration of strikingly beautiful insects takes place in Costa Rica. Green uranias (*Urania fulgens*), diurnal moths that in their shape and size closely resemble swallowtail butterflies, gather in their thousands on the Osa Peninsula to begin a mass flight to the Caribbean lowlands. At the same time populations from Guatemala commence their migration across Central America toward Colombia. The migration lasts for a few months, during which time hundreds of these moths can be seen festooned on individual trees, feeding on the nectar of their flowers.

The reason for the migration is related to the host plant on which caterpillars of this species feed. In Costa Rica the favorite host plant of urania caterpillars is a large rainforest canopy liana, *Omphalea diantra*. This plant, a member of the family Euphorbiacea, is capable of producing toxic compounds to protect itself from

herbivores such as caterpillars. The level of the toxins in the plant tissue depends on the amount of damage inflicted by the insects feeding on it. Since these secondary compounds are costly to produce, the plant will only make them in large quantities if the damage is persistent over long periods of time. After a few years of being attacked by urania caterpillars, the toxins in the leaves accumulate to the point that the insects are no longer able to cope with them. At that point, female uranias stop laying eggs on the *Omphalea diantra* plants on which they themselves developed, and all the moths set out to look for another "naïve" population of plants that have yet to accumulate high levels of toxins in their leaves. After a few years, once the level of toxins

drops to the level tolerated by urania caterpillars, the plants are recolonized again by the moths.

The toxins ingested by urania caterpillars give them effective protection against predators. This protection is passed onto the adults, who advertise their unpalatability with black and metallic green coloration. Interestingly, a nearly identical combination of colors is present in the poison dart frog *Dendrobates auratus*, a species equally well protected from predators by alkaloid toxins in its skin. Since uranias and dart frogs frequently occur in the same area, it is possible that both benefit from the resemblance to each other, thus presenting another example of Müllerian mimicry.

Glasswing butterfly (*Ithomia patilla*)

Tiger moth (*Hylaruga* sp.)

Another possible example of Müllerian mimicry occurs between glasswing butterflies (Nymphalidae, tribe Ithomiini) and tiger moths of the genus *Hyalurga*. Glasswing caterpillars feed on plants of the family Solanaceae, rich in toxic secondary compounds (pyrrolizidines), and sequester them for their own protection. Adults of some species inherit this protection, while in other species males feed on nectar and pollen of toxic plants to regain the protection lost in the transition to adulthood. These toxins also give males the precursors needed to produce the sex pheromones that attract females. During mating the males pass some of their protective toxicity to females, and there is evidence that females preferentially select partners with the highest levels of protective alkaloids.

Hylaruga tiger moths are likely similarly protected by alkaloids sequestered from plants that they feed on while in the caterpillar stage. By assuming the appearance of glasswing butterflies, widely recognized by predators as unpalatable, the moths benefit from the protection it gives them and at the same time their own toxicity reinforces the warning signals sent by the glasswings.

The toxic tiger moth *Idalus crinis* displays a classic combination of aposematic colors.

Pseudatteria leopardina is a diurnal moth of the family Tortricidae. While most of its relatives are drab and cryptic, it is very colorful. Its conspicuous coloration and slow, fluttering flight pattern indicate that it might be highly toxic, although more research is needed to confirm this notion, as it has been little studied.

An aggregation of shield bug nymphs (Scutelleridae).

Shield bug nymphs (Scutelleridae)

A behavior often found in toxic, aposematically colored insects is a tendency to aggregate in groups, especially in their juvenile, larval, or nymphal stages. There are several reasons for this behavior. First, predators are more likely to learn to associate noxiousness with a particular visual signal if they are presented with multiple copies of the same pattern. At the same time, individual insects in the group gain the advantage of being immediately avoided by a predator that just had an unpleasant encounter with another individual in the group. And last, clustering in a group reduces the number of individual predators the population of prey is exposed to. In other words, if the same number of individuals were scattered, then the chance of each individual encountering a predator would be much higher.

This clustering behavior can often be seen among aposematic nymphs of shield and stink bugs (families Scutelleridae and Pentatomidae). These insects protect themselves by expelling compounds that smell and taste bad from dorsal glands on the abdomen (adult shield-backed bugs have additional glands on the thorax). The compounds are either extracted from plants or synthesized *de novo* by the insects.

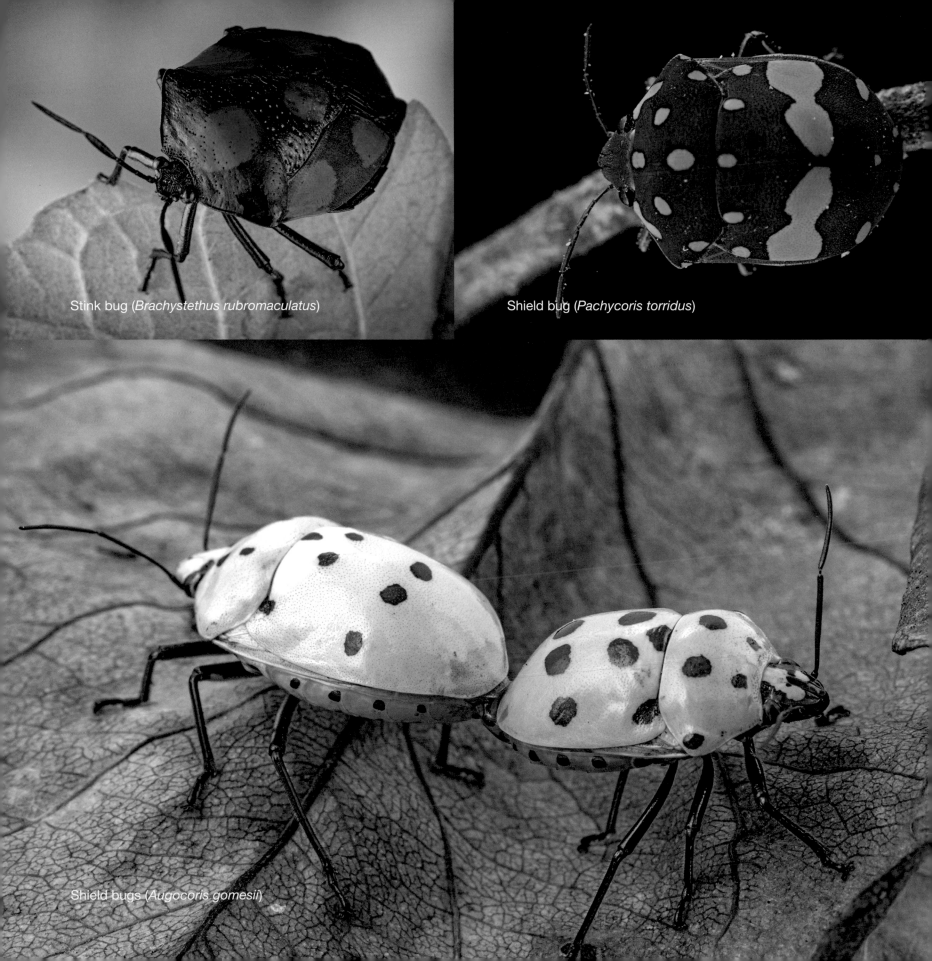

Stink bug (*Brachystethus rubromaculatus*)

Shield bug (*Pachycoris torridus*)

Shield bugs (*Augocoris gomesii*)

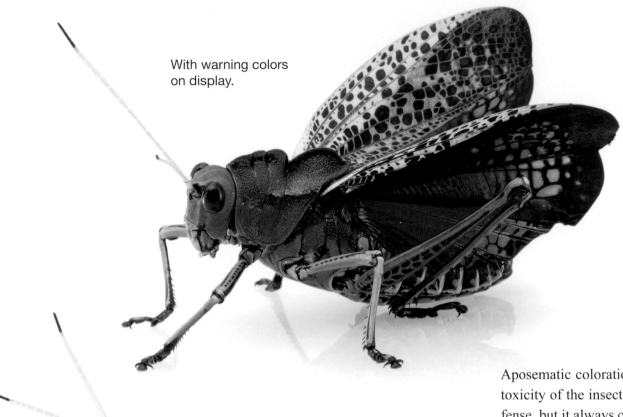

With warning colors on display.

In cryptic pose.

Aposematic coloration that warns predators of the toxicity of the insect's body is a very effective defense, but it always carries with it the risk of injury from a naive predator who has never before attacked a chemically protected species. An alternative approach is to show cryptic coloration generally and only display warning colors if an attacker is clearly intent on causing harm. This strategy is adopted by species whose toxicity varies depending on diet. Lubber grasshoppers (*Taeniopoda*), for example, are generalist herbivores that feed on a variety of plants, some that are rich in toxic secondary compounds and some that are not. This means that certain individuals or populations will have chemical defenses from the sequestered toxins, while others won't. Lubber grasshoppers thus rely primarily on cryptic coloration to avoid detection from predators. If attacked, they fan out their brightly red hind wings, which they normally keep hidden. They may also expel a foul smelling hemolymph to amplify the visual warning signal.

Caterpillar of swallowtail butterfly (*Papilio thoas*) displaying its osmeteria in reaction to danger.

Caterpillars of swallowtail butterflies (family Papilionidae) often resemble bird droppings in their shape and coloration, which allows them to avoid detection by many predators. But when attacked, the caterpillars change their appearance by suddenly growing a pair of large, brightly colored "horns" that emit a foul smelling substance. These are osmeteria, unique defense organs found only in swallowtail larvae. Once extended, they disperse butyric acids and other smelly compounds. These chemicals are not sequestered from plants on which the caterpillars feed but are synthesized *de novo* by the insects. Interestingly, younger, smaller caterpillars of certain species produce terpenoids, compounds that effectively protect them from ants. Once they reach full size—and are no longer under the threat of being attacked by ants—they switch to producing butyric acids, which are more effective in protecting them from vertebrate predators.

Leaf beetle (*Alurnus ornatus*)

Planthopper (cf. *Baleja* sp.)

Although biologists have made great strides within the last 50 years in understanding the ecology and systematics of Costa Rican insects, there are still countless species, often common and conspicuous, about which we know surprisingly little. The leaf beetle *Alurnus ornatus* is a large insect, commonly found feeding on leaves of palm trees. Its gaudy coloration and slow movements suggest that it is chemically protected, but direct evidence is lacking. Leaf beetles are known to both sequester toxic compounds from the plants on which they feed as larvae and to synthesize toxic compounds *de novo* as adults. A piece of circumstantial evidence is the existence of harmless insects, such as this unidentified planthopper (family Cicadellidae), that appear to be mimicking this leaf beetle.

Stick insect (unidentified Anisomorphini).

Walking sticks are known to synthesize *de novo* protective compounds, including toxic anisomorphal, which may cause temporary blindness in mammals. In Costa Rica, several species of the tribe Anisomorphini, whose members are known to produce these toxins, live at higher elevations in the Talamanca Range. They are often black and flightless, with highly shortened bright yellow wings that warn of their chemical defenses. These insects seem to carry an unusual number of parasitic mites (family Trombidiidae). The bright red coloration of the mites makes the insect more visible to predators, but it is possible that their presence also enhances its aposematic signal, thus offsetting the negative effect of parasitism and enhancing the walking stick's chances of survival.

Examples of Aposematically Colored and Chemically Protected Costa Rican Insects

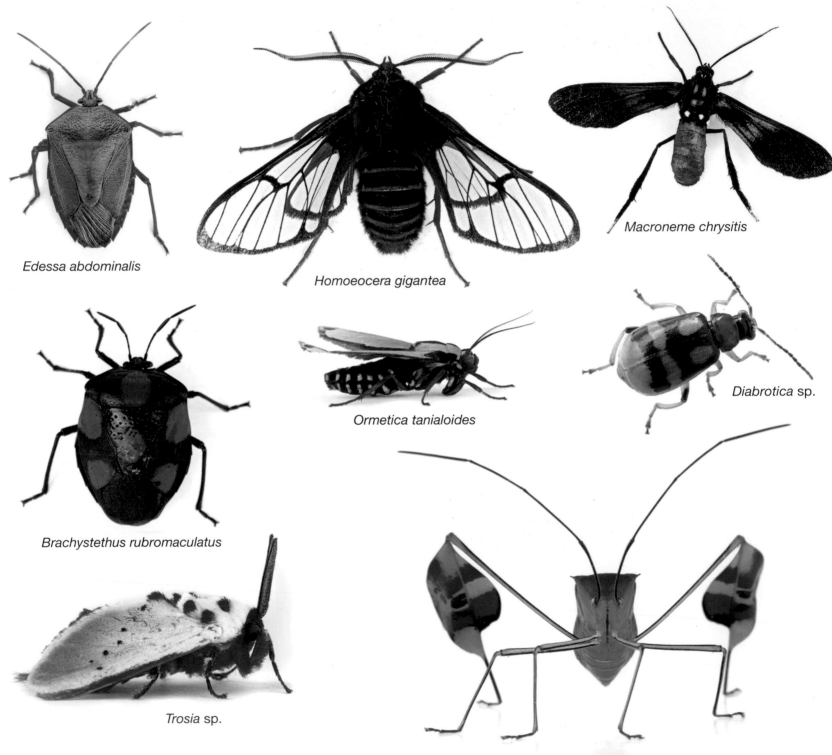

Edessa abdominalis

Homoeocera gigantea

Macroneme chrysitis

Brachystethus rubromaculatus

Ormetica tanialoides

Diabrotica sp.

Trosia sp.

Anisoscelis flavolineata

6
Sounds of the Forest

Zammara smaragdina

7:00 a.m. The sun rose over an hour ago and birds have been singing since its rays first began to illuminate the horizon. But only now has the forest warmed up sufficiently to induce cold-blooded singers to begin their acoustic activity. Along a narrow path in the Atlantic rainforest, a thick carpet of fallen leaves hides a mystery caller whose buzzy chirping emanates from a tangle of dead branches at the base of a large tree. Although very insect-like, the call's source is actually a tiny strawberry dart frog (*Oophaga humilis*). At this early hour most singing insects are still in hiding, silent. But soon a droning sound begins to filter down from high in the canopy, quickly building up to a loud cacophony that in some places overwhelms all other natural sounds. Sounding more like machinery—and lacking any melodic qualities—these are the calls of cicadas (order Auchenorrhyncha). Often heard but rarely seen, cicadas are large, winged insects that resemble oversized flies. They are very cryptic, and even species that tend to sing from lower tree branches are incredibly difficult to spot. One indication of their presence, apart from the whelming noise they produce, is a light shower of honeydew droplets that can be felt if you stand directly under a tree on which these animals are feeding.

Cicadas have a unique mechanism of sound production. In the majority of singing insects, sound is produced by rubbing one part of the body against another. This mechanism, known as stridulation, is responsible for the calls of crickets and grasshoppers, the defensive squeaking of longhorn

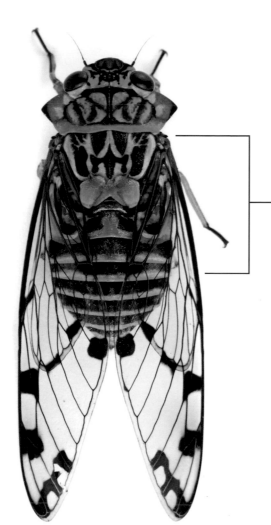

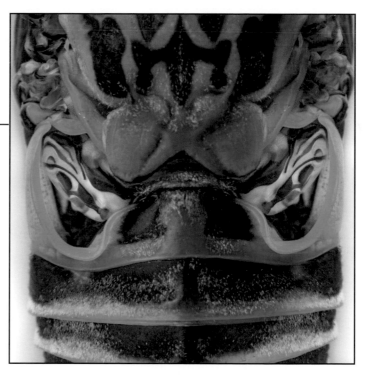

Close-up of tymbals of male cicada.

beetles, and the aggressive buzzing produced by some ants. But cicadas have evolved a mechanism that works on a very different principle. Lifting the wings of a male cicada reveals a pair of rounded openings on the posterior part of the thorax. Within each opening sits a large, white membrane adorned with a series of darker ribs. These are known as tymbals. To create a sound signal, powerful muscles in the thorax buckle the ribs, causing the tymbal to bend, followed by an instantaneous relaxation that makes the tymbal snap back into its original position. This cycle repeats itself 300–400 times per second, producing loud clicks that to our ears meld into a steady, buzzing sound. The left and right tymbals can buckle synchronously or independently of each other, resulting in songs of various frequencies and patterns.

It is only the males that sing, using their advertisement call in an attempt to attract a suitable female. A female indicates her selection from the pool of singing applicants by making a short click with a flick of her wings, and the selected male is then allowed to approach.

Oscillogram of the call of *Zammara smaragdina.*

Hamadryas februa

12:00 p.m. By noon the heat is at its peak. Few animals are active, preferring to hide in the merciful shade under the forest canopy. Only male cicadas seem to relish the high temperatures, continuing their ear-piercing competition for the attention of females. But another mysterious singer or, more precisely, noise maker, joins the din of their chorus. Every now and then a loud crackling sound can be heard among the trees, though the source of the sound is difficult to pinpoint. But with a little patience, you discover that the maker of the sound is, astonishingly, a butterfly. Cracker butterflies (*Hamadryas*) are highly territorial and protective of their mates and food (sap oozing from trees and rotting fruit). Males engage in spiraling chases between trees, producing loud crackling as they fly. The sound is likely produced to intimidate rivals, to chase away other insects competing for access to food, and to startle predators. The crackling sound also facilitates the formation of groups of reproductive adults and in some species males use it to impress females.

First discovered and described by Charles Darwin during his voyage to South America, the mechanism of sound production in *Hamadryas* is still not fully understood. It is clear that the clicks are made with the wings, but entomologists do not agree on *how* the clicks are made. A widely adopted hypothesis assumes that the sound is produced by thickened veins on both front wings; in flight, the veins hit each other as they meet in the upswing above the butterfly's body. More recent experiments, however, have demonstrated that *Hamadryas* can produce the sound with a single wing, negating the percussive explanation. While the precise mechanism is still unknown, it is likely that it involves a thickened wing vein that buckles and snaps, thus producing the crackling signal.

But what good is sound if nobody can hear it? The ability of *Hamadryas* butterflies to hear has long been a mystery. If the only purpose of their sound is to scare off predators and to ward off insects attracted to sap or fruits, then the cracker butterflies themselves don't need to be able to hear the sounds they produce. But if the sound is used in interactions with members of the same species, whether rivals or mates, then they must have the capacity to hear. It appears that *Hamadryas* perceive sound through the earlike Vogel's organ at the base of the wing. Interestingly, it is possible that Vogel's organ evolved when the ancestors of *Hamadryas*, once nocturnal moths, used it to avoid nocturnal predators such as bats. In most butterflies, the Vogel's organ—a "degenerate bat detector"—is non-functional, but here it has found another use.

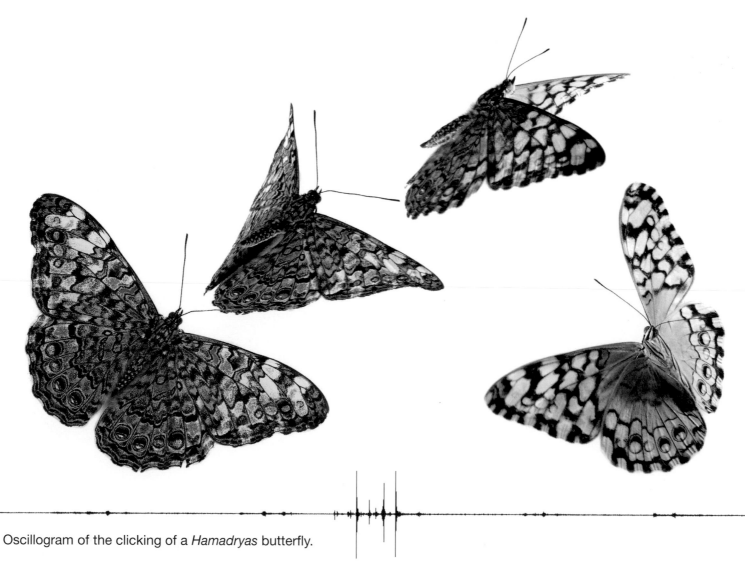

Oscillogram of the clicking of a *Hamadryas* butterfly.

5:00 p.m. As the sun prepares to disappear behind the horizon and the air starts to cool, a new suite of singers gets ready to begin their acoustic displays. With clockwork precision, almost exactly at 5:00 p.m., great tinamous, large elusive birds of the rainforest floor, begin their mournful yet strangely beautiful whistling. A few minutes later the first tree crickets join the birds in this short, early evening, concert. Species of the tree cricket genus *Anaxipha* start calling at very precise times, some at exactly 5:15 p.m., and continue for only a few minutes, after which they remain silent until the following evening. As the night approaches, more and more species of crickets start to sing and before nightfall the forest is vibrating with songs of thousands of insects.

Anaxipha sp.

Treehopper (*Guayaquila* sp.)

But beneath this loud display of musical prowess, a multitude of signals inaudible to our ears are also being transmitted. Air is not the only medium that can transmit sound waves. Acoustic signals can travel through liquids and solid matter as easily as through the air, and, for certain sound frequencies, tree trunks and stems of plants are a much better substrate to send and receive acoustic messages.

Until recently scientists assumed that the only insects that communicate using sound waves are those that humans can hear. Among those, cicadas were the only members of the order Auchenorrhyncha that were thought to use sound to communicate. But new technologies of sound recording have revealed a previously hidden world of busy interchanges of signals that take place out of our earshot—and that of predators.

Treehoppers (family Membracidae) have a repertoire of calls that puts to shame even the most loquacious of birds. Each species produces unique signals that can transmit messages of desire, fear, aggression, and loneliness. Mother treehoppers maintain constant contact with their offspring by sending a steady stream of signals through the branches of the trees on which they feed. At the same time males "talk" to each other, asserting the ownership of a particularly good spot on their chosen plant, while the females listen to the calls that attempt to demonstrate the male's ability to produce superior offspring. When approached by a predator, treehoppers often make buzzing noises that help repel the attacker. Although these insects produce some of the most beautiful sounds in the animal kingdom, they remain unknown to all but a handful of researchers.

An ant (*Camponotus* sp.) tending a treehopper (*Harmonides* sp.).

Treehopper (*Bocydium* sp.)

Treehopper (*Thrasymedes variata*)

Treehopper (*Umbelligerus* sp.)

You can perceive treehopper calls only if you have a device that taps into the sounds they transmit through the substrate on which they sit. They use a number of sound production mechanisms. Some species employ structures very similar to the tymbals found in cicadas; some rub elements of their abdomen against each other; and yet others drum with their abdomen against the substrate. Such substrate-borne communication has its advantages and disadvantages. The most obvious advantage is that treehoppers' signals cannot be detected by anybody who is not sitting on the same branch or stem. Predators, who often use the signals of singing insects to locate their prey, cannot zero in on treehoppers because they cannot hear them. But this turns into a problem for male treehoppers trying to attract females living on a different plant; thus the males must fly from plant to plant in search of females.

Treehopper songs that are transmitted through substrate may have another function. These insects, like many of their relatives in the insect order Auchenorrhyncha, produce copious amounts of sweet honeydew that is eagerly sought by ants. Many ant species treat treehoppers in the same way we treat our cattle, protecting them (by warding off predators) and "milking" their honeydew (see photo on p.143). Indeed, there is some evidence that some species of treehopper use special songs to attract ants.

As signal-recording technology continues to improve, entomologists have begun to understand that acoustic signaling behavior is present in nearly all other groups of the Auchenorrhyncha. Lantern bugs (family Fulgoridae) produce low frequency buzzing of yet unknown function and substrate communication has been documented in a virtually every group that has been tested for such behavior.

Lantern bug (*Scaralis neotropicalis*)

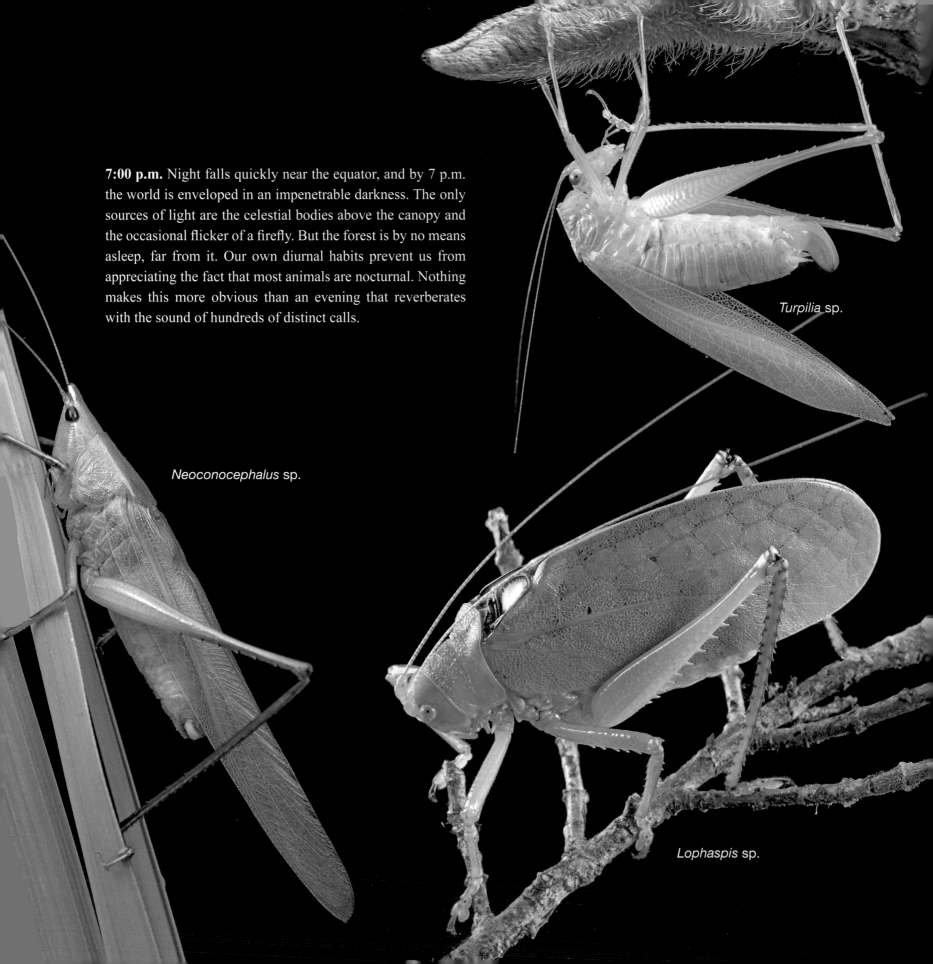

7:00 p.m. Night falls quickly near the equator, and by 7 p.m. the world is enveloped in an impenetrable darkness. The only sources of light are the celestial bodies above the canopy and the occasional flicker of a firefly. But the forest is by no means asleep, far from it. Our own diurnal habits prevent us from appreciating the fact that most animals are nocturnal. Nothing makes this more obvious than an evening that reverberates with the sound of hundreds of distinct calls.

Turpilia sp.

Neoconocephalus sp.

Lophaspis sp.

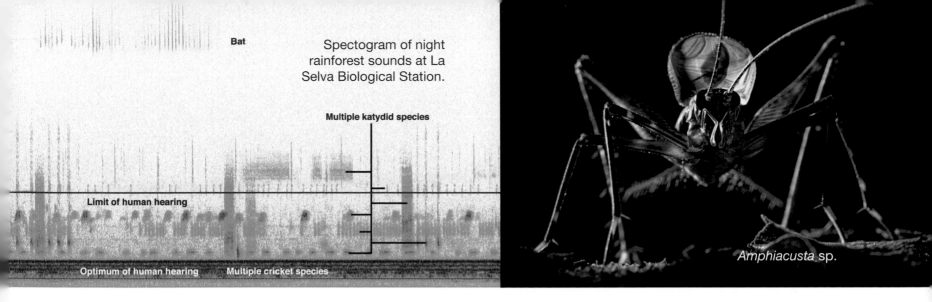

Bat

Spectogram of night rainforest sounds at La Selva Biological Station.

Multiple katydid species

Limit of human hearing

Optimum of human hearing   Multiple cricket species

Amphiacusta sp.

While some insect species begin calling early in the evening, katydids and crickets usually do not start calling in full force until they feel completely safe in the darkness. In Costa Rica nearly 500 species of katydid and cricket fill the night with songs that vary widely in volume, duration, and frequency. Some produce short, buzzing notes, while others make melodious, almost bell-like calls. Katydid calls tend to be less musical, often producing sounds similar to the drone of cicadas, though generally shorter in duration. The songs of crickets—more melodious than those of katydids—are usually delivered in long, uninterrupted bouts. Both katydids and crickets produce sound by rubbing their front wings against each other. These two groups, along with frogs during the rainy season, dominate the evening soundscape in Costa Rica. But what appears to be a peaceful chorus is in fact a vigorous contest for a slice of the available sound frequency spectrum.

If a call is to effectively attract mating partners and fend off potential rivals, it must be heard and understood by all members of the species—and be distinguishable from the calls of other species. For this reason, each species of singing insects produces a unique signal that is recognized as conspecific only by other individuals in its population. Such unique signals prevent females from mating with males of a different species, and stop males from engaging in unnecessary battles over non-conspecific females.

In addition to producing a unique call—the more unique, the better—each singing insect aims to maximize the efficiency with which it is produced. Depending on the environment, certain frequencies carry farther than others. If an insect is prevented from using the optimal frequency because another individual is already using it, it must either call louder, call at another time, or switch to a different frequency. This competition for airwaves is remarkably similar to the competition between media companies for radio and TV waves. The end result of this competition is a nearly complete saturation of the soundscape. Human ears can only appreciate a small slice of the wall of sound that fills the night sky. While our ears are theoretically capable of hearing frequencies up to 20 kHz (1 Hz = one sound wavelength per second), most of us are attuned to much lower frequencies. Add to this the fact that anyone older than 30 effectively loses the ability to hear anything above 15 kHz, and we can see that we miss a lot of the activity that fills the airwaves.

Most animals sing to attract mates. But prospective mates are not the only individuals attentive to mating calls, with many predators (and parasites) using the calls of katydids and crickets to locate and devour them. Thus, males of singing insects run a much greater risk of death from predators than females. From the point of view of the survival of a given species, this is perfectly fine since it is females who contribute more toward the continuation of their species, as

they produce eggs and, in some instance, take care of the young. But in some katydids this situation is reversed.

While in no animal species can males lay eggs or give birth, in some species they do provide females with the material and energy for creating the next generation. This is exactly what happens in some katydid species. During mating the male not only inseminates the female but also

A male *Calamoptera* sp. delivers a nuptial gift known as a spermatophylax (second from far left), a valuable package of carbohydrates, proteins, and minerals. Once the partners separate, the female eats the spermatophylax (right), while the male must rest and forage for at least a few days to regain his strength and recover the lost weight.

delivers a nuptial gift known as a spermatophylax, a valuable package of carbohydrates, proteins, and minerals. The spermatophylax nourishes the female, providing her the rarest and most difficult to obtain food ingredients, and it sustains her as she produces a new clutch of eggs. In such cases, the male's investment in offspring can be staggeringly high. In some species of katydids, for example, the male loses 40% of his body mass in a single mating. Once the partners separate, the female eats the spermatophylax, while the male must rest and forage for at least a few days to regain his strength and recover the lost weight. In katydid species in which the male's parental investment matches—or exceeds—that of the female, there is often a reversal of roles doing courtship, in which it is the female who seeks to win the male's favor, and he is liable to reject her if she is deemed an unsuitable mate.

10:00 p.m. After the initial explosion of insect sounds as night begins, the forest appears to quiet down. A few katydids and crickets continue to sing but otherwise the air is mostly silent. Or is it? Human ears, adapted to detecting low frequencies, are simply unable to detect the dramatic acoustic war fought out by insects, right above our heads.

Bats are fast, merciless predators of insects. Their weapon of choice is echolocation, a highly sophisticated use of sound signals that allows them to detect even the smallest object in absolute darkness. Using their throat and tongue, bats emit extremely high frequency pulses of sound that bounce off objects and return to the bat's ears with a brief delay. The length of the delay allows the mammal to judge the distance to the object, while the infinitesimally small difference between the sound perception in the left versus the right ear helps them judge the shape of the object. A similar technique is used by submarine sonars to detect objects and map underwater terrain. But unlike human-made sonars, which use very low sound frequencies, bat echolocation employs some of the highest pitched sounds existing in nature, some of which exceed our ability to hear by an order of magnitude. Some bat species produce echolocation calls in the 200 kHz range, or ten times the maximum sensitivity of human ears. Not surprisingly, we are oblivious to most of these animals' aerial activity.

But insects, with their far more sensitive ears, are fully aware of the presence of bats. Some, like katydids and crickets, simply hide in dense vegetation or under bark to avoid detection and to make themselves more difficult to extract. But insects that need to fly at night must resort to other strategies to avoid the echolocating hunters. Moths are the principal targets of insectivorous bats. Being smaller, weaker, and slower flyers, they would stand no chance against a pursuing bat; but they can hear bats, and this gives them an advantage. Upon

detecting an approaching bat, many moths simply fold their wings and drop from the sky straight to the ground, like a rock. Some moths employ a craftier defense; as a bat approaches its prey, the rate of echolocation signals increases to produce a more detailed picture of the insect. And this is what some moths wait for—as soon as the bat is only a few inches away, the moth blasts its own series of ultra-high-frequency pulses, effectively jamming the bat's sonar and making it miss its target.

Tiger moths are known for their toxicity, which they advertise with their bright aposematic coloration. Alas, this strategy is useless against a nocturnal, color-blind hunter. Fortunately, the insect's unpalatability can be conveyed through sound, and many tiger moths emit loud warning clicks that have the same effect as the warning colors. On hearing these clicks, the bats determine that the insect is not one to be trifled with and abandon the pursuit.

Nocturnal tiger moth (*Hylisidota*)

7

Living Together

Gregarious blattodeans (*Capucina*)

We humans pride ourselves on a number of remarkable, world-altering achievements: agriculture and the domestication of animals; the use of electricity; and the formation of large, cooperative societies whose members have the ability to convey complex messages to others and where individuals act altruistically toward each other. All of these developments, we presume, are unique steps in our evolutionary history that place us above the rest of the animal world. And yet, as we are discovering, we are not the pioneers that we once thought, but rather copycats who have plagiarized ideas and solutions already employed by other organisms for millions of years. With the exception of the use of electricity, which many fish use very effectively to detect and overpower their prey, insects have beat us to many innovations that seemed to have made our species so exceptional.

Social behavior—the ability of individuals of the same species to live together and often participate in activities that benefit other members of the group—has evolved independently in several groups of unrelated arthropods. Within insects, blattodeans were likely the first to begin forming complex societies. Modern species of this order display a gamut of behaviors that includes juveniles staying with their mother for a prolonged period of time; aggregations of individuals that share the same shelter and food resources; and large colonies where individuals of different generations live together and benefit from each other's presence. But about 150 million years ago, a lineage of blattodeans began a process that lead to the creation of the first complex societies, in which each member has a responsibility that benefits the colony as a whole. We know them now as termites, a subset of blattodeans that traditionally has been placed in its own order, the Isoptera.

Termite societies are eusocial, or truly social. In a eusocial society each member has a clearly predetermined role, from which it does not deviate throughout its entire life. The group's multiple generations live together, and reproductive rights of individuals are strictly controlled. This stringent division of labor, based on a system of distinct castes, means that most termites lead a life of celibacy and daily drudgery—all for the good of the queen and the king, the only individuals in the colony capable of reproduction. The sterile, wingless, and blind worker caste, which constitutes the majority of individuals in the termite society, will spend their life taking care of the queen's brood, building and repairing the nest, and collecting food. Although genetically identical to their ruling parents, the development of the reproductive organs and wings of the workers is suppressed by pheromones produced by the queen. Similarly, the soldier caste includes sterile, wingless, and

Termite workers.

blind individuals, but their main role is to defend the colony. Soldiers usually have large, heavily sclerotized (hardened) heads. In some termites the weapon used by soldiers is a pair of powerful, extremely sharp mandibles. Such species can easily cut through the skin of even the largest attackers.

Instead of sharp mandibles, soldiers of the subfamily Nasutitermitinae have an enlarged head that houses a big frontal gland that is topped with a long, forward facing nozzle (nasus). The frontal gland secretes various chemicals, mostly terpenes, that are very effective at repelling ants—the principal enemies of termites—and even large predators such as tamandua anteaters. Nasutitermite soldiers usually lack functional mandibles and must thus be fed, mouth-to-mouth, by workers who masticate the food before delivering

it. One species in this subfamily that does have mandibles is the spectacular *Rhynchotermes perarmatus*, whose soldiers have a long nasus and also sport giant, hook-like mandibles.

Though sterile, workers and soldiers of termites can be genetically male or female, which stands in contrast to ants, bees, and other social Hymenoptera, whose non-reproductive castes are always female.

Usually each termite colony has just one queen and one king, but in some species colonies sometimes have secondary queens, which become reproductive after the establishment of the colony by the primary queen. Members of the reproductive caste are much larger than workers or soldiers. Every year the colony produces a number of reproductive

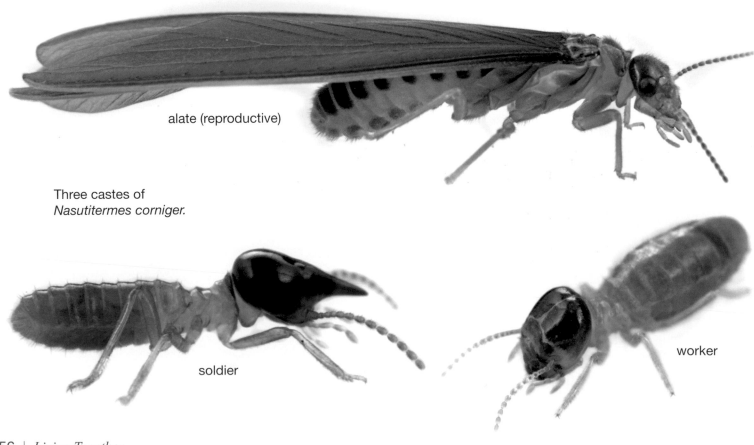

alate (reproductive)

Three castes of
Nasutitermes corniger.

soldier

worker

Termites foraging in leaf litter.

individuals that leave it to start their own colonies. In the initial stage the reproductives are equipped with two pairs of large wings and are known as alates. Following their nuptial flight, the alates pair up, break off their own wings, and begin to build a new colony. After founding the colony, the queen transforms from a normal-looking termite into a giant egg-laying machine. Her abdomen swells up to become hundreds of times larger that that of a typical termite, and becomes filled with ovaries that produce a constant stream of eggs. Such females, known as physogastric queens, live a remarkably long time; there is evidence that in some species the queen may reach the age of 50, perhaps even 70, years.

Termites, who are among the few organisms capable of digesting cellulose, feed primarily on wood. They are able to digest cellulose, in part because of the symbiotic bacteria and protozoans in their gut that produce cellulase, a digestive enzyme capable of breaking it down. Additionally, the termites themselves have evolved their own cellulitic enzymes. Termites are also able to digest lignin, a polymer that makes wood hard and

Rhynchotermes perarmatus

Nasutitermes corniger

A termite nest (*Nasutitermes corniger*).

long-lasting and that is generally difficult to break down; they can break down the lignin possibly thanks to symbiotic fungi that live in the gut of these insects, although little is known about this symbiotic relationship. Termites' ability to digest and thus recycle wood makes them critically important members of tropical ecosystems. We tend to think of them as pests, but without them the world would be sinking under ever accumulating layers of dead branches and leaves.

Most species found in Costa Rica build their nests on trees or other elevated places to avoid being flooded during the rainy season. The nest is made of "carton," finely chewed wood that is mixed with feces to make it more pliable and sticky. Termite nests are a sought-after commodity, and other animals—including birds, for whom a carton nest makes a great site to lay eggs—often attempt to evict the rightful tenants.

Orchid bees (*Euglossa* sp.)

The evolution of eusociality did not stop with the termites. Members of the order Hymenoptera embarked on the same path at least 100 million years ago, as evidenced by amber fossils showing that ants already had a caste system during the Cretaceous Period. Similar to what can be witnessed among the blattodeans and termites, members of the order Hymenoptera display a range of social behaviors; some species are solitary, some gregarious, and yet others display truly social behaviors.

Interestingly, sociality is known only in those groups that have developed a stinger (Aculeata) and is absent in other members of this group of insects. Within stinging Hymenoptera, bees (family Apidae) exhibit a full range of social behaviors, from small, non-eusocial colonies to fully eusocial ones.

Orchid bees (*Euglossa*) are beautiful, metallic-surfaced (in hues of green and blue) bees, whose females sometimes form small colonies without a distinct division of labor. In species that raise their young communally, each female contributes to the construction of the nest but forages and feeds her own young independently of other members of the colony. Interestingly, in some cases the number of females in the colony exceeds the number of the brood, indicating that some females may abstain from reproducing in order to perform other functions.

Male orchid bees are unusual among insects in that they collect perfumes from flowers, mostly orchids. As the bees collect these fragrant, volatile compounds, pollen sticks to them and they unwittingly disperse it to other flowers. The males store the perfumes in the enlarged tibia of the hind legs, later using them to create pheromones that will attract females.

Paper wasps (subfamily Polistinae) form small, eusocial colonies that display a distinct division of labor. Although all members of the colony, which are invariably females, are capable of reproduction, only the dominant female ("foundress") lays eggs. She aggressively prevents other females from reproducing, turning them into de facto workers. In the absence of the dominant female, however, other members of the colony may start laying eggs.

One common trait of eusocial insect colonies is mouth-to-mouth feeding of adults by other adults. Known as trophallaxis, this behavior is believed to strengthen the bonds between members of the colony and help establish a hierarchy of dominance. In paper wasps, subjugate females frequently feed the dominant females.

Paper wasps (*Polistes* sp.)

Paper wasps (*Polybia* sp.) feeding each other. This behavior, known as trophallaxis, strengthens the bonds between individuals.

Stingless bees (*Tetragonisca angustula*) are common in Costa Rica, and their nests can be found in both natural habitats and around human-made structures. Their colonies are very large (up to 10,000 individuals) but extremely well hidden in hollowed out trees, rock cavities, or even within the walls of houses. The entrance to the colony, which is sometimes sealed at night, is marked by a long wax tube that is guarded day and night by larger workers. Despite being stingless, these insects are protected by their well-hidden nest. They can also bite and release formic acid on the attacker.

Stingless bees are eusocial, generally with only the queen laying the eggs that develop into workers. Some workers, however, may occasionally lay eggs, though they are unfertilized, which means that they will develop into males. Sex determination in the Hymenoptera is haplo-diploid. Fertilized eggs—those with two sets of chromosomes (diploidy), one from the male and one from the female—develop into females, whereas unfertilized eggs—with only one set of chromosomes (haploidy)—develop into males. The result of this sex determination mechanism means that in a eusocial colony all female workers are more closely related to each other than they are to their own mother (since they only share half of their genes with her). This is considered one of the reasons for the emergence of altruistic behavior within eusocial societies, in which individuals will often sacrifice their life to protect other members of the colony.

Stingless bees (*Tetragonisca angustula*), here and below.

Acromyrmex coronatus

Ants are the very apex of eusociality. These highly modified social wasps are members of the family Formicidae. Dating back to the Cretaceous Period, at least 100 million years ago, ants have become the dominant form of life in many terrestrial habitats. Despite the small size of an individual ant, their abundance makes them responsible for 15–20% of the animal biomass of the globe! In the tropics, where ants are most abundant, they make up at least 25% of the animal biomass and within lowland rainforests their biomass can be four times higher than that of mammals. This is particularly evident in Costa Rica, where one of the first animals a visitor is likely to encounter is a leafcutter ant, usually seen in a seemingly endless column on the forest floor.

Leafcutter ants, members of the genera *Atta* and *Acromyrmex*, are an excellent example of insects' superiority over humans

Atta cephalotes

when it comes to the invention of agriculture. The first Europeans to see leafcutter ants assumed that the bits of vegetation that these ants were carrying were used to build thatched roofs to protect the colony from tropical downpours. It was not until 1874 that the famous naturalist Thomas Belt realized the true use of the leaves. Leafcutter ants are expert gardeners, and the greenery is used as a substrate for growing nutritious fungi (*Leucoagaricus gongylophorus*), the main source of food and water for the ants and their larvae. The fungi have been truly domesticated by the ants and are no longer able to reproduce and grow on their own. Until recently this symbiosis was thought to involve just these two organisms, but two additional players are now known to be a part of this relationship. Another fungus, a parasitic member of the genus *Escovopsis*, often invades, and can attack and destroy the ants' garden. This parasite usually occurs in

Atta cephalotes

Acromyrmex worker carrying another worker; the white powder on its body is the antibiotic-producing bacteria *Pseudonocardia*.

An underground fungal garden of *Sericomyrmex amabilis*.

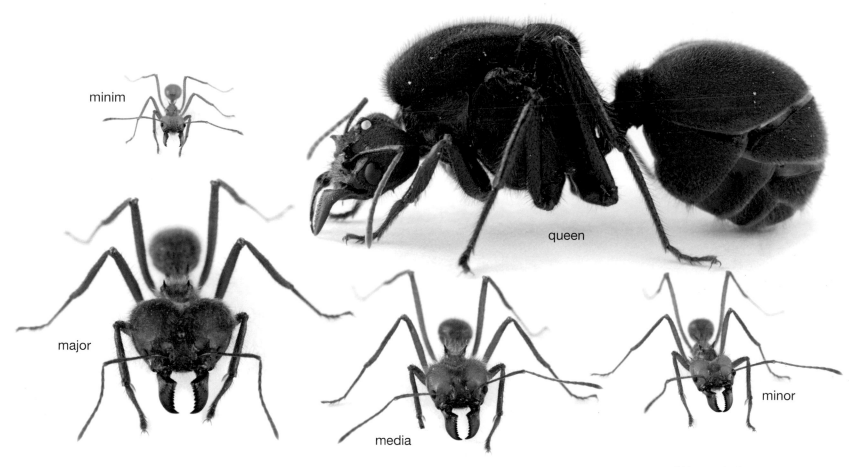

minim

queen

major

media

minor

Atta cephalotes castes.

low numbers, but when the health of the garden is compromised, it becomes virulent, overgrowing the entire fungal crop. Fortunately, ants have a powerful weapon against it, an antibiotic-producing bacteria of the genus *Pseudonocardia*, which they carry on their bodies and use to control the growth of the parasitic fungus. (Incidentally, a study of these bacteria in the nests of Costa Rican fungus-growing ants has recently resulted in the discovery of a new, promising antibiotic, selvamicin.)

Leafcutter ants display polyethism, a division of labor often accompanied by extreme differences in body size. In *Atta cephalotes*, a common species in Costa Rica, the colony has four different types of workers, each responsible for a different task, in addition to the giant queen (males die shortly after mating). The largest workers are the majors, big-headed individuals that act as soldiers to defend the colony. Occasionally they participate in other tasks, such as carrying particularly heavy pieces of vegetation back to the nest. Mediae belong to a caste responsible for most of the leaf cutting; minors help with carrying leaves and, alongside the majors, defend the column of foraging ants from potential attackers; the smallest individuals are known as minims, and their main function is to take care of the brood and the fungal gardens. They can also be seen riding on top of leaf fragments carried by larger workers. In this case their role is to protect the worker from deadly parasitic flies (Phoridae) that attempt to lay eggs on the body of the leaf-carrying ant.

Bivouac of *Eciton hamatum*.

The rainforest floor is a dangerous place if you are a small, flightless animal. Army ants (*Eciton*) can overpower and kill virtually any organism that they encounter during their foraging raids. But unlike other predatory ants, these raids do not start from and return to a permanent nest. This is because army ants do not build nests, despite the fact that their colonies can be enormous, with up to 600,000 individuals. Instead, their life is divided into two phases. During the nomadic phase, which lasts about two weeks, the entire colony, including the queen, is constantly on the move during

Eciton mexicanum

the day, stopping at a new location every night. A raiding column may include 200,000 individuals and stretch for hundreds of meters. Each day the raid goes into a different direction to avoid areas where the ants have already foraged. While streaming across the forest floor some members of the column assume the role of living bridges, allowing other members of the column to stream over holes and other obstacles. In order to make the progression of the column as fast and smooth as possible, on encountering a branch or a hole in the ground, multiple individuals stretch their bodies to create a platform on which other members can quickly pass. To facilitate this behavior, the tarsi (feet) of army ants come equipped with sharp hooks that make gripping the substrate, and each other, much easier.

In the statary phase, which lasts about three weeks, the entire colony congregates in one spot, usually at the base of a tree or under a fallen log, and forms a bivouac, a large basket—made of their own bodies—that protects the queen and her eggs, which lie concealed in the middle of the bivouac.

Workers of *Eciton burchellii parvispinum* forming a living bridge over a gap in the ant column's path.

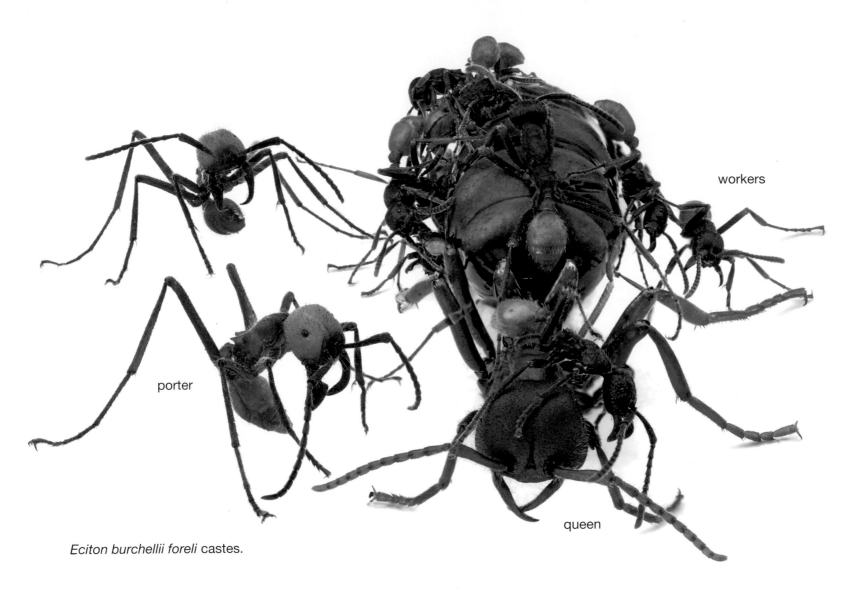

porter

workers

queen

Eciton burchellii foreli castes.

To collect food for the colony, a smaller proportion of its members set out on raids every other day. The bivouac remains in place until the new eggs have hatched and a new generation of adults emerges from the pupae. After this the bivouac reorganizes itself into a massive column, a process that can take up to eight hours, and the colony sets out on a new nomadic phase.

Like leafcutter ants, army ants display a remarkable degree of polymorphism. The workers are fairly typical-looking ants, and their main job is to find and kill prey. The porter caste includes larger ants with strong, enlarged mandibles. As the name implies, their function is to carry pieces of dismembered prey killed by the workers. And then there are soldiers, whose massive sickle-shaped mandibles are designed to impale any enemy unwise enough to cross path with the army ants. Their aggression and tenacity is well known across Latin America. Certain tribes of Amerindians use army ant soldiers as living sutures to close and heal wounds—once the mandibles of the ant have penetrated human skin, they will never let go. In contrast to the other castes, the queen is a giant, several times longer than even the largest of the workers. Her only job is to

Eciton burchellii foreli soldier.

produce eggs, and she is entirely dependent on her children for food and defense. When the colony is on the move, the queen is completely surrounded by a thick mat of workers and soldiers who will defend her to the death.

Army ants are a deadly force in the Costa Rican rainforest, and animals flee in all directions at the first sign of their approach. Other animals take advantage of the chaos caused by the army ants to pick out insect, frogs, and lizards escaping the raid. Antbirds specialize in following army ant raids to catch insects flushed out by the ants and also steal larger items already captured by them, while parasitic flies of the genus *Stylogaster* go only after blattodeans running for their lives.

Queen of *Azteca* ant.

Cecropia tree.

Müllerian bodies on *Cecropia* tree.

Worker (*Azteca* sp.) tending treehoppers.

Cecropia spp. is a common pioneering tree, frequently seen along the roads and in open gaps of the Costa Rican rainforest. The favorite food of sloths, it is also one of the species favored by leafcutter ants. Thankfully, another group of ants, genus *Azteca*, protect the trees from complete defoliation. *Azteca* ants have evolved a mutualistic relationship with *Cecropia* trees by living in hollowed-out stems of the plant and chasing away herbivores, including leafcutter ants, that might endanger the tree. The tree also benefits from nitrogen found in ants' refuse and their dead, decomposing bodies. In addition to providing housing for the insects, *Cecropia* repays for these services by producing Müllerian bodies, nutritious packets of glycogen, a rich source of carbohydrates for ants.

In the dry woodlands of northern Costa Rica, you can observe the remarkable relationship between bullhorn acacias (genus *Vachellia*) and ants of the genus *Pseudomyrmex*. A long coexistence between these two organisms has produced a system in which the plant depends on the insects for protection from herbivores and other plants competing for the same space, while the ants are incapable of living anywhere else. The ants regularly patrol the tree, warding off or killing any insect trying to feed on the leaves of the acacia. They also remove seedlings of other plants from around the trunk of the acacia, eliminating the possibility of the host tree becoming overshadowed by other trees. They will even prune branches of neighboring trees if they threaten to overshadow the host. In fact, acacias do very poorly and are easily outcompeted by other plants if the ants are absent. In return, the ants reap a number of benefits. At the base of the petiole of each leaf are small, elevated organs known as extrafloral nectaries that regularly exude droplets of fluid rich in sugar and amino acids. These are eagerly collected by the ants. Young leaves also produce Beltian bodies—nutritious, protein- and lipid-rich packets that the ants use to feed their larvae. The tree itself is a home for the ants. Large, sharp spines on its branches are hollow, making perfect chambers to raise the brood and provide shelter for the queen and workers.

Pseudomyrmex sp. collecting nectar from extrafloral nectaries.

Pseudomyrmex flavicornis collecting Beltian bodies from young acacia leaves.

Ants (*Camponotus* sp.)
tending gregarious
treehoppers (*Harmonides* sp.)

A rich social life and sophisticated agricultural practices are not the only things that insects invented long before humans did. The practice of animal husbandry—the care and cultivation of another species of animal for food—has been a part of ants' evolutionary history for millions of years. The mutualistic relationship of ants with many species of aphids, treehoppers, and other groups of the order Auchenorrhyncha is well documented. Ants protect these insects from attackers and in return are rewarded with nourishing droplets of honeydew, a sugar-rich byproduct of the Auchenorrhyncha diet of plant juices. The relationships between these groups is so close that they have developed a system of signals understood by both sides. Treehoppers respond to gentle tapping of the ant's antennae by producing a fresh droplet of honeydew, whereas ants are attuned to acoustic signals transmitted through the stems that seem to lead them to the treehoppers' location.

In addition to termites, ants, bees and their relatives, eusocial behavior is also known in ambrosia beetles, small insects living under the bark of trees who, like leafcutter ants, have evolved fungal agriculture. Less sophisticated social behavior is found in other groups of insects, including crickets, true bugs, caterpillars, and many others. Among arthropods, social and cooperative behavior is not restricted to insects.

Ant (*Pheidole* sp.) collecting honeydew from treehoppers (*Bolbonota xalapensis*).

Azteca sp. worker tending treehoppers.

Pseudoscorpions (*Paratemnoides elongatus*) with prey, a large male of army ant (*Eciton* sp.).

Social spiders (*Anelosimus* sp.)

Sociality is not an attribute usually associated with arachnids. In fact, cannibalism is common among these animals. But false scorpions have evolved social behaviors that allow them to hunt prey much larger than any individual could overpower if working alone. False scorpions' venom glands open at the tips of their "pincers," or pedipalps. The venom is not very fast acting, especially if the victim is large. The prey must be held in place long enough for the venom to act and paralyze it. Members of the species *Paratemnoides elongatus* form large family groups under the bark of rainforest trees,

sticking out their grasping pedipalps to catch unsuspecting passersby. Many individuals, young and old, hold the prey, while at the same time injecting venom into its body, and soon the feeding can begin.

Among spiders, members of the genus *Anelosimus* form large family groups that not only build much larger webs than any individual spider could on its own but also display cooperative hunting behavior that allows them to overpower very large prey. The prey is then shared by all members of the group.

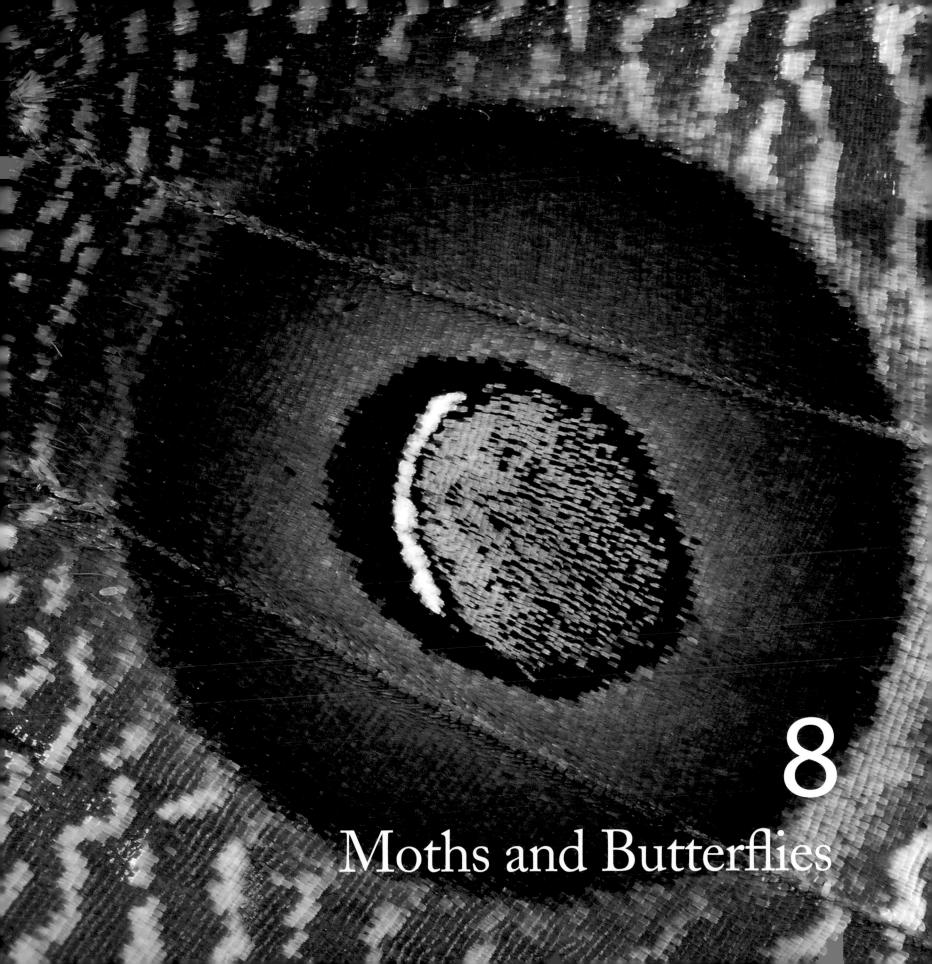

8

Moths and Butterflies

Unlike most of our mammalian cousins, who prefer to conduct their business at night, we are diurnal animals, and for this reason humanity has been on a quest to light up the darkness ever since we learned how to make fire. Today, our nights are brightly illuminated by artificial sources, a comfort to us but a disaster of cataclysmic proportions for the rest of the natural world. This becomes quite evident if we look at a streetlight or a lamp on the porch of a house and see that hundreds, perhaps thousands, of insects are drawn to their seductive glow. Many of them will never leave, circling the lights until they die of exhaustion or fall prey to bats or geckos. Those that remain on the walls beneath the lights will be picked off at dawn by birds. A few lucky ones will manage to escape, only to return the following night. It seems that insects are drawn to artificial lights like a moth to a flame.

Surprisingly, after hundreds of years of observing this behavior, scientists are still not entirely sure why insects display phototaxis, as this behavior is technically known. Several explanations have been proposed, none entirely convincing. The most widely accepted explanation is based on the fact that at night insects use celestial bodies for navigation. The light coming from the stars and moon is polarized (its waves travel only in one direction), making it possible for an insect to maintain a constant flight path in relation to the direction of the light waves. An artificial light radiates in all directions, confusing the insects and making them spiral ever closer to the source. Some insects may come to lights because they confuse them with openings

These moths (right) were photographed during a single night under a light bulb in front of a ranger station in Braulio Carillo National Park. They represent fewer than 15% of the moth species that were present under the lights that night.

Silk moth (*Copaxa syntheratoides*)

that indicate an unobstructed flight path. Another, rather far-fetched, theory suggests that artificial lights are confused with the sex pheromones of female insects, which sometimes give off a very faint glow.

Whatever the true explanation, on any given night human-made lights end up attracting untold trillions of insects, mainly moths, the most abundant and species-rich group of flying insects in almost any environment.

Moths are one of the largest groups of insects, second only to beetles. Over 160,000 species of moth have already been described, but this likely represents only slightly more than 50% of the actual number of species in the world. In Costa Rica alone, at least 15,000 species are estimated to be present; and, in some families of Costa Rican moths up to 98% of species remain undescribed by biologists. Moths and butterflies, the latter nothing more than a relatively small group of diurnal moths of the superfamily Papilionoidea, have achieved

Concealer moth (*Struthoscelis* sp.)

this staggering success thanks to a number of morphological, physiological, and behavioral innovations, chief among them their intimate, coevolutionary relationship with plants.

Several traits define moths, including the presence of fine, colorful scales that cover their wings and most of their body (the name of the order Lepidoptera comes from Greek words meaning "scaly wings"). Moth scales are modified setae, the insect equivalent of hair. They perhaps first evolved to

function as insulation to help the moths maintain a higher body temperature, a requirement for these fast flying insects. But scales have several other important functions. First and foremost, moth scales are an infinitely malleable canvas for the evolution of colors and shapes. No other group of insects approaches the chromatic diversity of moths. Color patterns created by the scales can be used to send warning messages of their carrier's unpalatability, to attract mates and scare off enemies, to distort the recognizable shapes of the moths'

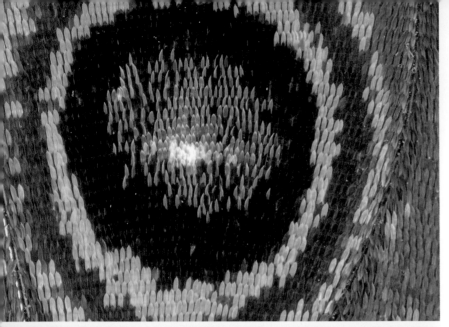

bodies, or to produce invisibility cloaks that allow them to disappear against almost any background. The coloration of moth scales is either pigment-based or structural. Melanins, uric acid, carotenoids, and a few other types of compounds are responsible for pigment-based coloration in moths, and such coloration tends to fade with age. Brilliantly blue and green markings of some tiger moths, *Urania*, and *Morpho* butterflies are the result of light scattering on the microscopically sculpted surface of individual scales. Such structural colors never fade.

Scales are also a defensive "weapon," and anybody who has ever held a moth will attest to the fact that the scales on their wings make these insects incredibly slippery. Scales are designed to dislodge at the slightest touch, making it difficult for a bird or other predator to get a good grip on the insect's body. A fall into a spider web, a death sentence for most insects, often results in the moth making a quick escape, as it leaves behind a thick coating of scales on the sticky silk strands. The slipperiness of scales also reduces the amount of friction caused by wings rubbing against each other, thereby reducing the amount of muscle power needed to propel the insect in flight. Female moths frequently use their scales to protect egg clutches from desiccation and predators, or to mark oviposition sites to indicate to other females that this particular spot has already been spoken for.

Scales on the wings of a *Morpho* butterfly (left). Scales on the underside of the wings display pigment-based coloration, while the metallic blue scales on the upper side show structural colors.

Morpho helenor

Glasswing butterflies (tribe Ithiomiini) are unusual in that they have greatly reduced amounts of scales on their wings. Their transparent wings allow them to virtually disappear in the shady understory of the forest but, if detected by a potential predator, their white and orange markings warn of the poisonous properties of their bodies, which they obtain by feeding on plants rich in toxic pyrrolizidine alkaloids.

Glasswing butterfly (*Oleria* sp.)

Glasswing butterfly (*Ithomia patilla*)

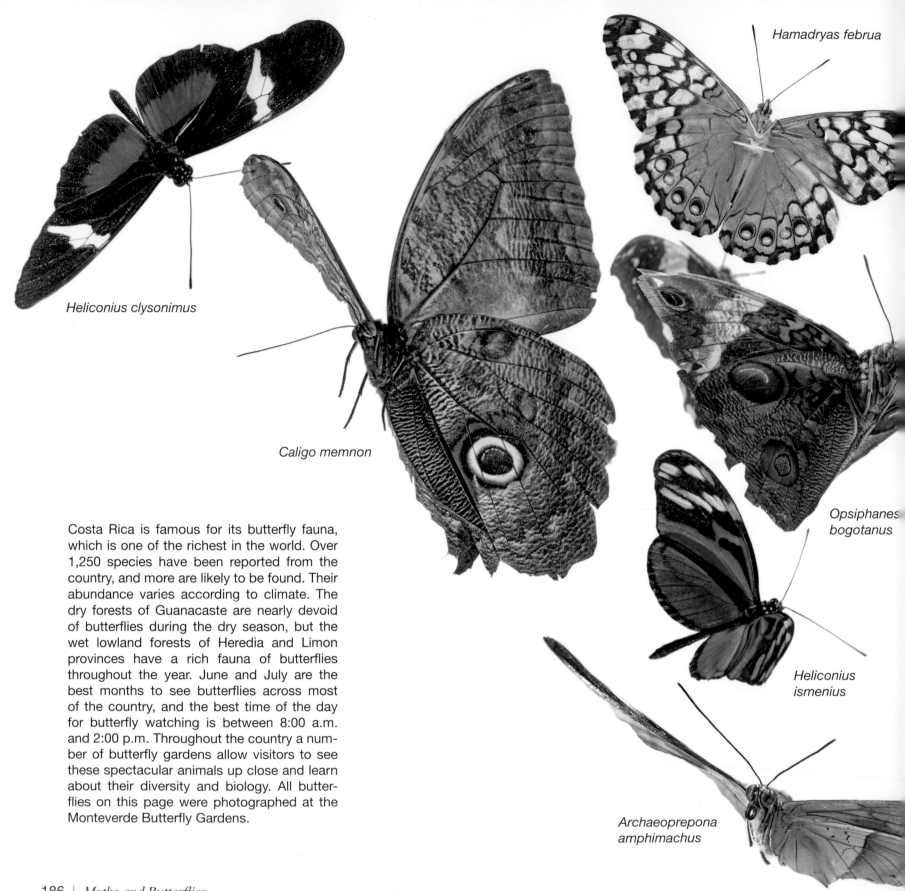

Heliconius clysonimus

Caligo memnon

Hamadryas februa

Opsiphanes bogotanus

Heliconius ismenius

Archaeoprepona amphimachus

Costa Rica is famous for its butterfly fauna, which is one of the richest in the world. Over 1,250 species have been reported from the country, and more are likely to be found. Their abundance varies according to climate. The dry forests of Guanacaste are nearly devoid of butterflies during the dry season, but the wet lowland forests of Heredia and Limon provinces have a rich fauna of butterflies throughout the year. June and July are the best months to see butterflies across most of the country, and the best time of the day for butterfly watching is between 8:00 a.m. and 2:00 p.m. Throughout the country a number of butterfly gardens allow visitors to see these spectacular animals up close and learn about their diversity and biology. All butterflies on this page were photographed at the Monteverde Butterfly Gardens.

Siproeta stelenes

Eueides isabella

Caligo memnon

Morpho helenor

Catonephele numilia

Laparus doris

Morpho helenor

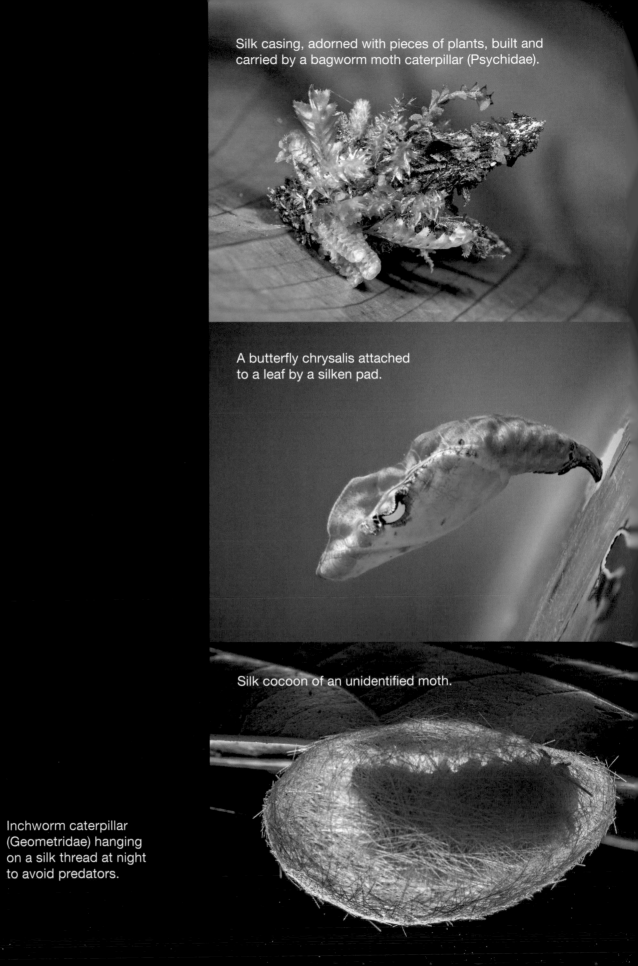

Silk casing, adorned with pieces of plants, built and carried by a bagworm moth caterpillar (Psychidae).

A butterfly chrysalis attached to a leaf by a silken pad.

Silk cocoon of an unidentified moth.

Inchworm caterpillar (Geometridae) hanging on a silk thread at night to avoid predators.

The evolutionary success of moths is also strongly tied to their ability to produce copious amounts of silk. This amazing polymer, whose tensile strength exceeds that of steel, has allowed moths to increase their mobility—and likelihood of survival. Very early on, our own species realized the value of silk; indeed, the Asian silk moth (*Bombyx mori*), whose single cocoon can yield an 800-meter-long uninterrupted strand of silk, was one of the first domesticated animals. But before a caterpillar uses its silk glands to spin a protective cocoon around itself, it often uses silk in very different ways. Newly emerged caterpillars, for example, use silk for "ballooning," to float through the air to new locations (a similar behavior is seen in young spiders). Older caterpillars of some moth species use a single, long strand of silk to suspend themselves at night from a branch to avoid marauding ants and other predators. Caterpillars of the families Psychidae and Tineidae build silken shelters adorned with bits of plants, lichens, and wood to protect their soft bodies. In some groups, caterpillars spin large communal nests made of silk that protect them from predators while they feed on leaves inside.

When the time comes for caterpillars to turn into a pupa, the caterpillars use silk to firmly anchor themselves to the substrate. Some then use their silk spinnerets, located on their mouthparts, to weave a waterproof, almost impenetrable, cocoon around themselves. Those species whose larval body is covered with defensive hairs and spines will often shed the hairs and spines and weave them into the fabric of the cocoon to increase its protective properties. Strangely, caterpillars of the family Yponomentidae spin beautiful, elaborate baskets that leave the pupa almost fully exposed rather than fully enclosing the pupa in a protective silken cocoon.

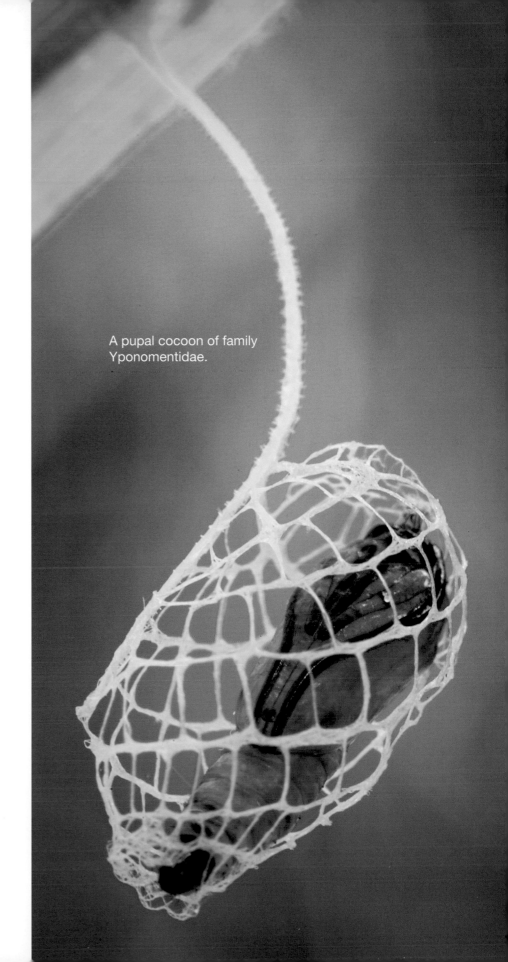

A pupal cocoon of family Yponomentidae.

Colobura dirce

Siproeta stelenes

Laparus doris

Dryas iulia

About 3 billion years ago, oxygen-producing cyanobacteria created a breathable atmosphere on our planet. As momentous as that event was, the coevolutionary relationship between insects and plants is probably the second most important biological event in the history of our planet. The explosive radiation of flowering plants in response to insect pollination and herbivory has resulted in the origin of nearly every terrestrial ecosystem that we now know. Our own civilization is a byproduct of this event, as without insects and modern plants, mammals—including primates—would never have been able to evolve as they did. Moths have played the leading role in this coevolutionary process.

Nearly everything about a moth's body—most notably the mouthparts—is designed for interacting with plants. A major evolutionary innovation was the appearance of the proboscis, a long, flexible tube that can be used to reach nectar at the bottom of even the longest flower corollas. A small group of moths still retains chewing mouthparts that allow them to grind pollen and the spores of ferns, but the majority of species carry an elaborate suction device that restricts

them to imbibing food in liquid form only. The proboscis can be conveniently coiled when not in use, reducing drag in flight and preventing the mouthparts from being damaged.

Coevolution between flowers and moths is beautifully illustrated by plant species that produce flowers with corollas too long to access by all insects other than a single species of moth. The most famous example comes from Madagascar, where a star-shaped orchid with an exceptionally long nectary prompted Charles Darwin to speculate about the probable existence of a moth with an extremely long proboscis. It wasn't until 20 years after Darwin's death that such a moth was indeed discovered. In Costa Rica the sphinx moth *Amphimoea walkeri* has a 28-cm-long proboscis, the longest of any moth, which allows it reach to the bottom of flowers inaccessible to other insects. The process of pollination by a specific moth species has allowed plants to limit the amount of reproductive cells they need to produce. Having pollen gathered by moths that preferentially feed on only one species of

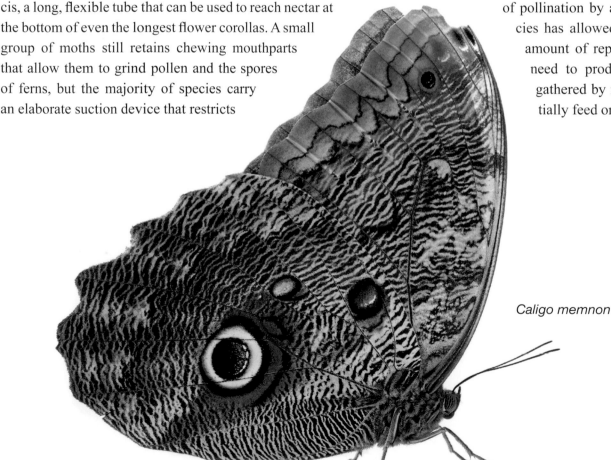

Caligo memnon

An owlet moth (Noctuidae) feeding on partially eaten fruit.

plant guarantees that the pollen will end up on another flower of the same species, carried there on the head and body of the moth. Plants that rely on wind pollination or pollination by non-specialized, "messy" pollinators (such as beetles) must produce copious amounts of pollen to increase their chances of fertilization, clearly not an optimal strategy. Although the sucking mouthparts of moths restricts them to a liquid diet, which is often poor in amino acids and other nutrients easily found in solid food, some species have managed to overcome this problem by producing digestive enzymes that allow them to feed on nutritious pollen. Such species collect clumps of pollen on their proboscis, and then slowly liquefy and drink it while resting on a branch.

Some moth species never visit flowers, obtaining their nutrients instead from other sources. It is common to see moths aggregating around rotting fruits, sap seeping from tree trunks, piles of dung, and even carcasses of dead animals. A few species specialize on drinking the sweat and tears of mammals (mostly to obtain sodium, an element scarce in plant tissues), and yet other species use their sharp, piercing proboscis to puncture the skin of animals to drink their blood.

Tithorea sp. Note clumps of pollen on the proboscis of the butterflies.

Laparus doris

Of course, the relationship between plants and moths is not always mutually beneficial. While most plants would not be able to exist without moths' reproductive services, moths are also the number one enemy of plants. Their larvae or caterpillars feed on the leaves and stems of plants, threatening their very survival. Plants defend themselves in a myriad of ways, mostly through the production of chemical defenses that make their tissues difficult to digest or downright deadly. Of course, over time moths almost invariably evolve counter-defenses that allow them to sidestep plants' chemical warfare, and the cycle begins anew. But one Costa Rican plant has evolved a different strategy. Butterflies of the genus *Heliconius* lay their eggs on passion vines (*Passiflora*), and their voracious caterpillars often cause tremendous damage to individual plants. But *Heliconius* females will not lay eggs on a plant if they know that another female has already done so, as a single plant cannot support the brood of more than one butterfly. Passion plants exploit this bias by marking their leaves with fake "butterfly eggs" of the same size and bright yellow color as the true eggs. Upon seeing these markings, the female usually abandons her plans to lay her eggs there.

Heliconius melpomene

Eggs of *Heliconius* butterfly and a false, egg-like marking on a passion flower (*Passiflora*) leaf.

Passion flower (*Passiflora coccinea*)

Photo Credits

All photographs in this book are by Piotr Naskrecki.

Front cover: Crayola katydid (*Moncheca elegans*)
Back cover: Leaf katydid (*Aegimia venarecta*)

p. iv: [clockwise from top left] Leopard moth (*Pantherodes pardalaria*), leaf beetle (*Alurnus ornatus*), dobsonfly (*Platyneuromus soror*), planthopper (undet. Cicadellidae), grasshopper (*Aidemona azteca*), treehopper (*Guayaquila* sp.)
p. v: Leaf beetle (*Diabrotica regalis*)
p. viii: Crayola katydid (*Moncheca elegans*)
p. 1: Tree mantis (*Liturgusa* sp.)
p. 17: Leaf katydid (*Orophus tessellatus*)
p. 53: Bullet ant (*Paraponera clavata*)
p. 67: Rhinoceros katydid (*Copiphora rhinoceros*)
p. 89: Leaf mantis (*Choeradodis rhombicollis*)
p. 113: Caterpillar of the sphinx moth (*Pseudosphinx tetrio*)
p. 137: Cricket (*Amphiacusta* sp.)
p. 153: Army ants (*Eciton burchelli*)
p. 177: Scales on the wing of the owl butterfly (*Caligo eurilochus*)
p. 196: Unidentified moth

Index

204

About the Author

Born in Poznan, Poland, Piotr Naskrecki is an entomologist, conservation biologist, and photographer. He received his Ph.D. in entomology from the University of Connecticut. Currently he is Associate Director of the Edward O. Wilson Biodiversity Laboratory at Gorongosa National Park, Mozambique, and Research Associate at the Museum of Comparative Zoology, Harvard University. Naskrecki is the author of *The Smaller Majority*, *Relics*, and numerous scientific papers and photographer of *A Window on Eternity* by Edward O. Wilson.